Handbook of
House Plants

The Authors

Dr. M. Kannan is presently working as the Professor and Head, Department of Floriculture and Landscaping, Horticultural College and Research Institute, Tamil Nadu Agricultural University, Coimbatore. He started his carrier in 1986 as Assistant Professor (Hort.) at Tamil Nadu Agricultural University, Yercaud. To his credit he has published more than 9 books, several technical bulletins, student manuals and book chapters related to horticulture and specifically to floriculture. During the tenure of his service spanning over 31 years, Dr. Kannan has been involved in UG/PG/Ph.D teaching, research on floriculture and many extension oriented activities. He is the recipient of 10 awards and is a member of various professional and academic societies. He was the Project Officer for the Tamil Nadu Precision Farming Project which sensitized several horticultural farmers of Tamil Nadu on hi-tech production of horticultural crops using drip and fertigation.

Dr. P. Ranchana is working as a Post Doctoral Fellow in Tamil Nadu Agricultural University, Coimbatore. She has published 6 international, 5 national and 30 popular articles, 6 manuals and 5 book chapters. She served as a Course Associate for UG and PG courses with respect to Floriculture and Landscaping. She is a life member of several professional societies. She use to handled classes for ODL students towards Floriculture and Landscaping aspects. She assisted in the conduct of 25 days training programmes for women entrepreneurs at the Department of Floriculture and Landscaping, TNAU, Coimbatore.

Dr. S. Vinodh is working as a Post Doctoral Fellow in Tamil Nadu Agricultural University, Coimbatore. He has published 5 international, 4 national and 15 popular articles, 2 books and 8 book chapters. He received 5 awards including INSPIRE - Fellowship for his Doctoral Research programme from DST, New Delhi. He assisted various training programmes for farmers and women entrepreneurs and ODL Programme conducted at the Department of Floriculture and Landscaping, TNAU, Coimbatore.

Handbook of
House Plants

M. Kannan

P. Ranchana

S. Vinodh

Department of Floriculture and Landscaping
Horticultural College and Research Institute
Tamil Nadu Agricultural University
Coimbatore

2018

Daya Publishing House®

A Division of

Astral International Pvt. Ltd.
New Delhi – 110 002

Cataloging in Publication Data--DK
Courtesy: D.K. Agencies (P) Ltd. <docinfo@dkagencies.com>

Kannan, M. (College teacher of floriculture and landscaping), **author.**
Handbook of house plants / M. Kannan, P. Ranchana, S. Vinodh.
    pages cm
    ISBN 9789387057258 (Int. Edn)

    1. House plants--Handbooks, manuals, etc.   I. Ranchana, P., author. II. Vinodh, S. (College teacher of floriculture and landscaping), author. III. Title. IV. Title: House plants.

    SB419.K36 2017   DDC 635.965   23

Published by   :   **Daya Publishing House®**
A Division of
Astral International Pvt. Ltd.
–ISO 9001:2015 Certified Company –
4736/23, Ansari Road, Darya Ganj
New Delhi-110 002
Ph. 011-43549197, 23278134
E-mail: info@astralint.com
Website: www.astralint.com

Preface

House plants are an ideal way to create attractive and restful settings while enhancing our sense of well being. They refer to a group of ornamental foliage and flowering plants, normally grown in containers and used to decorate the house interiors, verandahs, corridor, porch, stair, window sill *etc.*, In addition, houseplants can be a satisfying hobby and can help purify the air present in our homes. Indoor plants not only convert carbon dioxide to oxygen, but they also trap and able to absorb many pollutants. Many of these polluting chemical compounds are released into the air through a process called "off-gassing," from almost all the items which we use daily in our homes and offices.

The cultivation of house plants is by no means a difficult proposition. It is easy to achieve success, if one has the right knowledge about their habit and habitat, needs and susceptibilities, cultural practices and propagation techniques, the pests and diseases likely to cause damage, and their prevention and control. These information have been presented well in different chapters of this book.

In growing plants indoor, the people face some starting problems. Keeping this in view, some practical suggestions have been included for growing indoor plants successfully. Though numerous indoor plant species are available, it has been mentioned only those species that are most common and easily available from nursery outlets.

The publication of this book will be useful to both, professional botanists and horticulturists on the one hand, and amateur gardeners, nurserymen, plant lovers, architects and interior decorators on the other. At this juncture, we would like to thank all the staff and students of the Department of Floriculture and Landscaping, Horticultural College and Research Institute, TNAU, Coimbatore who extended cooperation during preparation of this book. We express our deep sense of gratitude to Astral International (P) Ltd. New Delhi, for their interact and support taking up this assignment and bringing out this publication in a nice manner.

M. Kannan
P. Ranchana
S. Vinodh

Contents

1

Introduction

Plants bring natural beauty to our living spaces. By creating texture and balance, a room or patio can be transformed into an environment that comforts us. Much of the scenic beauty of nature has been replaced by densely populated areas that sprawl for miles from urban centers. This visual pollution affects us all and leaves us with a longing for a closer connection with nature. We spend about 90 percent of our time indoors. The indoor environment is 5 to 10 times more polluted than the exterior.

House plants are an ideal way to create attractive and restful settings while enhancing our sense of well being. They refer to a group of ornamental foliage and flowering plants, normally grown in containers and used to decorate the house interiors, verandah, corridor, porch, stair, window sill *etc.*, In addition, houseplants can be a satisfying hobby and can help purify the air in our homes. Indoor plants not only convert carbon dioxide to oxygen, but they also trap and absorb many pollutants. Many of these chemical compounds, which are released into our air through a process called "off-gassing," come from everyday items present in our homes and offices.

Usually indoor plants are chosen for their ability to tolerate particular indoor conditions such as low light, high temperature, and low humidity. Plants that are able to grow well under these relatively adverse indoor environmental conditions make good indoor plants. If plants other than these rugged types are to be grown indoors, the environment must be altered to meet their needs or they must be put in a microclimate that is suited to their requirements.

History of House Plants

As far back as the time of the great pharaohs and the pyramids, the people of Ancient Egypt decorated their homes with plants. Displaying plants in containers was popular in Egyptian art. The Greeks and the Romans built their homes around a central atrium of containerized plants. And even though these plants were out in the open elements, this atrium was considered apart of the home. This was the fashion of such civilized societies. During the age of European discovery, the explorers who went to the New World found many new tropical treasures, many of which did not survive Europe's climate, so they were brought indoors. For centuries curious explorers and botanists, called plant hunters, searched the tropics for new and exciting plants to bring back to Europe. Since there was no suitable climate for these new treasures, plant houses needed to be constructed. The first of these was called an orangerie, like the one built at Versailles in France that housed the citrus and palms of king Louis XIV over the winter. Later conservatories and greenhouses like we know them today were developed. During the Victorian Age, exotic plants become extremely popular, but the transport of them was very difficult because of the lack of warmth, rooting media, and care on the ship ride home. On their long journeys, plants would become diseased or damaged and die until 1833, when Dr. Nathaniel Ward developed a glass case to transport the plants. These Wardian Cases were self sufficient on long journeys and became a fad in Europe for transporting exotic plants. Today we call these cases as terrariums. Also during this Victorian age, transition was in the air and technology changed the lives of the people. There was less free time in men's lives and the role of women changed in the home. Besides just taking care of the home, women chose horticulture as a hobby to become more active. This was at first just outdoor gardening, but there was a shift to gardening indoors in poor weather and the use of houseplants that we know today came into existence. This hobby took off and became extremely popular with both men and women and soon almost everyone had houseplants, especially in the cities. Before the 1940's the home environment was not suitable for tropical and subtropical plants, the popular foliage plants of today. After this time though, more precise control of indoor temperatures in both summer and winter greatly improved and there was no longer a need for conservatories and greenhouses to keep these plants over the winter.

Importance of House Plants

a) Control the Indoor Humidity

Plants release moisture vapour as part of the process of photosynthesis as well as its respiratory process. Because of this, it increases humidity of the air. Plants would release around 97 percent of the water they take in. If we want to increase humidity of a room,one can add more plants and put them together.

When humidity level is too low, individuals are more likely to develop viral infections; when humidity is too high, vulnerability to other disease may get increase. Plants tend to control humidity to within the optimum range for human health.

b) Easier Breathing

When the plants undergo photosynthesis, they absorb carbon-di-oxide and release oxygen. This mean that there is increased oxygen levels when there are plants indoors. During night time, they absorb oxygen and release carbon-di-oxide. Hence, the plants which grown indoors would be placed outside during evening hours. But there are certain plants which could absorb carbon-di-oxide and release oxygen during night time. Those plants will be placed in the bed room to get a peaceful sleep; they are orchids, succulents and epiphytic bromeliads. As a result of which concentration and productivity of a person may get increase to 10 – 15%.

c) Clean and Purify the Indoor Air

With the modern climate-controlled and air-tight spaces what we have these days, it is easier to trap volatile organic compounds (VOCs) which may be harmful to our health. Adding plants to the home will help to lessen these harmful gases. NASA says that houseplants can remove 87% of toxins from the air. These include gases like formaldehyde from cigarette smoke, rugs, grocery bags, and vinyl, benzene and trichloroethylene that are found in man-made fibers, solvent, sink, and paint. Plants will absorb these gases and pull these contaminants into the soil.

d) Reduce Stress

Plants can calm the heart rate and lower blood pressure. It can also reduce muscle tension that is related to stress. It could also decrease headaches, fatigue, sore throats, colds, coughs, and flu-like symptoms. It has a great impact to improve the health of human being. House plants make us feel comfortable and will even influence us to feel positive.

e) Sharpen Focus and Creativity

Studies show that when plants are placed in an indoor space, may be it a home, office, class room or in workplaces, there is an observed increase in productivity as a result of being able to focus on work. People also tend to be more creative because of having greens in the home.

f) Reduce Noise Pollution

It helps to reduce noise pollution. This happens basically because its stems, leaves, branches, wood, and other parts absorb sound.

g) Sets Higher Mood Levels

A home with more plants tends to be happier with less stress and fatigue. An environment with plants make the people would really be happy and cheerful because they would feel better in terms of health and the aura with plants around.

h) Soften the Look

Adding plants will create a contrast to hard surfaces. That is why; mostly the interior designing is concentrated by placing plants near the brick walls or sharp corners. Making the area appear softer will result into a more comfortable feeling and well-being in the spaces.

i) Improved Aesthetics

Be creative while choosing plants for indoors. Placing right plants for the right place is most important.

j) Other Benefits of Interior Plants

- Increased positive feelings and reduced feelings of anxiety, anger and sadness.
- Reduction of sound levels
- Reduction of stress levels
- Control of humidity to within the optimum levels for human health
- Cooling effect
- Absorption of carbon dioxide and emission of oxygen refreshing the air
- Improved concentration levels leading to improved productivity particularly with those working with computers
- Reduction of absenteeism in the workplace
- Faster recovery from mental tiredness
- Interiors feel spacious and clean
- People prefer to occupy rooms that contain plants
- Improved image - interiors are perceived as "more expensive"

Poisonous House Plants

Though house plants is blessed with several advantages, some of them are proved to be more beneficial in addition to its poisonous nature by causing illness if eaten, and others can cause skin irritation and should be handled with caution.

Children are unlikely to eat house plants, but some brightly colored fruit may seem tempting. Pet animals like cats and some dogs may play with or chew on plants. The toxic sap in poisonous house plants tastes extremely bitter and can cause a burning sensation in the mouth, so a pet is unlikely to play with them long.

Hence caution should be taken while pruning or repotting these poisonous house plants by keeping hands away from eyes and mouth, and to wash the hands thoroughly afterwards. It's better to wear gloves while handling or avoiding the following poisonous plants in home where children and pets are present.

Botanical name	Common name	Symptoms
Dieffenbachia	Dumb Cane	Sap causes painful swelling of mouth and throat, as well as vocal loss if eaten
Anthurium species	Anthurium	Leaves cause severe burning in mouth and skin irritation
Spathiphyllum	Peace Lily	Sap causes severe burning in mouth and skin irritation
Syngonium podophyllum	Arrowhead Plant	Sap is toxic and can cause skin irritation
Asparagus densiflorus	Asparagus Fern	Poisonous berries
Epipremnum aureum	Pothos or Devil's Ivy	Non-lethal, but causes burning sensation in mouth
Monstera deliciosa	Swiss Cheese Plant	Leaves cause severe burning in mouth if eaten
Philodendron species	Philodendron	Skin irritation
Schefflera species	Schefflera	Burning in mouth; skin irritation
Zantedeschia species	Calla Lily	All parts are poisonous, especially rhizomes
Chrysanthemum morifolium	Chrysanthemum	Leaves are poisonous if eaten and cause skin irritation
Codiaeum variegatum	Croton	Poisonous sap
Euphorbia milii	Crown of Thorns	Sap causes irritation in mouth and eyes
Hedera helix	English Ivy	Leaves are poisonous if eaten; sap can cause skin rash
Cycas revoluta	Sago Palm	All parts are poisonous; the seeds contain the highest amount of toxin
Zamioculcas zamiifolia	ZZ plant	All parts are poisonous

2

Ornamnetal Plants for Home Gardening

Types of Plants Used for Home Gardening

- Foliage plants
- Flowering plants
- Palms
- Conifers
- Climbers and creepers
- Ferns
- Cacti
- Succulents
- Flowering trees
- Annuals
- Spring/ bulbous ornamentals
- Shrubs
- Moisture loving plants
- Ornamental grasses & bamboos

Indoors plants for ornamental foliage

Common Name	Family	Botanical Name	Special Features
Acorus	Araceae	*Acorus calamus*	-
Aglaonema	Araceae	*Aglaonema angustifolia*	Very short stature
		Aglaonema crispum	Large pointed leathery leaves, slightly grayish green
		Aglaonema modestum	Chinese evergreen thrives in poorly lighted place, found to grow in bottles filled with only water
		Aglaonema pseudobracteatum	Deep green leaves variegated light green & yellow centre. very attractive & hardy
		Aglaonema 'Silver Queen'	Leathery lanceolate leaves with greenish silvery blotches on deep green background. Very bushy with a large clump
Alocasia	Araceae	*Alocasia argyraea*	Leaves dark green with silver sheen
		Alocasia cuprea	Leaves wide, purple beneath, dark metallic green above
		Alocasia indica	Very tall, green
		Alocasia indica var. *Metallica*	Deep purple leaves with metallic sheen
		Alocasia indica var. *Varigata*	Leaves and stalks mottled white with violet patches
		Alocasia 'Hilo Beauty'	Very attractive, thick leaves nicely marked with irregular areas of coloured, translucent blotches
		Alocasia 'Korthalsii'	Leaves long, deep olive green with white mid rib and veins, purple underneath

Contd...

Common Name	Family	Botanical Name	Special Features
Alpinia	Zingiberaceae	*Alpinia allughas*	Flowers red
		Alpinia rafflesiana	Flowers small with red tipped golden
		Alpinia speciosa	Indian shell flower (with spots & stripes of white, purple & yellow)
Alternanthera	Amaranthaceae	*Alternanthera amoena*	Leaves green veined with orange red blotches
		Alternanthera var. Amabilis	Leaves orange – scarlet
		Alternanthera bettickiana	Leaves blotched with shades of yellow and red
		Alternanthera versicolor	Leaves coppery red
Ananas	Bromeliaceae	*Ananas comosum variegatus*	-
Anthurium	Araceae	*Anthurium bakeri*	-
		Anthurium crassinervum	Lamina elliptic to oblanceolate
		Anthurium crystallinum	Leaves broadly ovate, enlarged green leaves, ribs bold, silvery
		Anthurium digitatum	A climbing type
		Anthurium magnificum	Leaves white & green
		Anthurium veitchii	Leaves evergreen with corrugated margins
Aralia	Araliaceae	*Aralia* sp.	-

Contd...

Common Name	Family	Botanical Name	Special Features
Asparagus	Liliaceae	*Asparagus densiflorus* 'Myers'	Plume like shoot like foxtail. the needle like foliage is deep green
		Asparagus densiflorus 'Sprengeri'	Stems are semi-prostrate, which grows to a height of 30cm. Branches are spreading over a diameter of 90cm. Wiry shoots bear cluster of bright green cladode. True leaves are modified into spines
		Asparagus plumosus	Feather triangular leaves are extremely beautiful, that spread horizontally. Stems are climbing with dark green colour. At nodes aerial roots develop as in bamboo. A leaf is tied to a button–hole flower
		Asparagus racemosus	Woody climber with heavy fasciculated roots. The massed cladode look very beautiful
Aspidistra	Liliaceae	*Aspidistra elatior*	Parlour palm, cast-iron plant
Beaucarnea (Ponytail palm)	Asparagaceae	*Beaucarnea recurvata* (Syn – *Nolina recurvata*)	Evergreen perennial growing to 15 feet 6 inches (4.72 m) with a noticeable expanded caudex, for the purpose of storing water inside
Bignonia	Bignoniaceae	*Bignonia masoniana*	Iron cross. The large roundish, firm green leaves are marked with palmately bands in the center
		Bignonia rex 'Silver Queen'	Painted – leaf begonia, leaves flat, soft silvery grey exception along the veins in centre & margin are metallic green
Calathea	Marantaceae	*Calathea insignis*	Silvery grey with bright green blotches linear, erect with wavy margin underside purple
		Calathea lindeniana	Dark green with emerald green zone, underside maroon
		Calathea ornata	Leaves dark green with pink (or) cream lines, dark purple beneath
		Calathea zebrina	Deep velvety green leaves yellow – green lateral veins

Contd...

Common Name	Family	Botanical Name	Special Features
Chlorophytum	Liliaceae	*Chlorophytum comosum*	-
		Chlorophytum elatum	-
		Chlorophytum elatum var. 'Variegatum'	-
Crotons	Euphorbiaceae	*Codiaeum variegatum* var. 'Andreanum'	Yellow
		Codiaeum variegatum var. 'Chelsonii'	Cream, green & red
		Codiaeum variegatum var. 'Tricolor'	Cream, green & gold
		Codiaeum variegatum var. 'pictum'	Large broad leaves with yellow, red & green variegation
		Codiaeum variegatum var. 'Bangkok'	Leaves roundish with yellow margins
		Codiaeum variegatum var. 'Excellent'	Dwarf with attractive, lobed foliage
		Codiaeum variegatum var. 'Iceton'	Small leaves having cream, pink, green & copper variegation
		Codiaeum variegatum var. 'Philip Geduldig'	Compact plant with broad lamina
		Codiaeum variegatum var. 'Trilobed'	Trilobed leaves variegated yellow & greens
		Codiaeum variegatum var. 'Indian Prince'	Very long slender leaves of screw – shaped with green, yellow & crimson variegation

Contd...

Common Name	Family	Botanical Name	Special Features
Coleus	Labiatae	*Coleus blumei*	Very dwarf plant of different colour variegation
		Coleus frederici	Annual (or) biennial
Cordyline	Liliaceae	*Cordyline australis* 'Purpurea'	Stems long, slender terminated with a rosette of linear leaves of purple colour
		Cordyline australis 'Cuprea'	Leaves are copper
		Cordyline australis 'Doucetii'	Pink stripes along the margin of leaves. Green portion may have white streaks
		Cordyline terminalis	Leaves long, broad. Tender leaves are bright reddish pink. Older leaves are coppery green.
		Cordyline terminalis 'Angusta'	Leaves long, slender, colour coppery green. Red margins underneath purple
		Cordyline terminalis 'Bicolor'	Leaves green with pink stripes along the margins
		Cordyline terminalis 'Madame Eugene Andre'	Bushy with broad, coppery green leaves having red margin. Old leaves drop after the emergence of flowers

Indoor plants with aesthetic flowers

Common Name	Family	Botanical Name	Special Features
Anthurium	Araceae	*Anthurium andreanum* var. rhodochlorum	Spathe salmon red with green extremities
		Anthurium andreanum var. giganteum	Large salmon red spathes
		Anthurium andreanum var. alba	Spathes white
		Anthurium andreanum var. roseum	Spathes shining
Aphelandra	Acanthaceae	*Aphelandra aurantiaca*	Flower head orange
		Aphelandra pectinata	Scarlet
		Aphelandra squarrosa	Flower head yellow with greenish – yellow bracts
		Aphelandra tetragona	Scarlet
Clerodendrum	Verbenaceae	*Clerodendrum thompsoniae* (Syn. *C. balfouri)*	Bright crimson flower with clear white bracts
Columnea	Gesneriaceae	*Columnea banksii*	Scarlet
		Columnea gloriosa	Scarlet & yellow
		Columnea gloriosa var. *purpurescens*	Scarlet & Yellow
		Columnea microphylla	Scarlet & yellow
Episcia	Gesneriaceae	*Episcia* sp.	Scarlet
Haemanthus	Amaryllidaceae	*Haemanthus katherinae*	Scarlet on separate stalks forming a dense head
		Haemanthus magnificus	Large head of orange scarlet flowers with golden stamens
Hedychium	Zingiberaceae	*Hedychium coronarium*	Plant resembling canna, flowers are white, strongly scented
Heliconia	Heliconiaceae	*Heliconia metallica*	Flowers red enclosed in narrow boat-shaped green bracts
		Heliconia striata	Flowers are long and orange
		Heliconia bihai	Flowers greenish yellow, bracts boat shaped, burnt red with yellow tip, the edges apple green to yellow

Contd...

Common Name	Family	Botanical Name	Special Features
Day Liliy	Liliaceae	*Hemerocallis aurantiaca*	Japanese day lilies – Flowers orange yellow
		Hemarocallis citrina	Flowers lemon yellow, slightly
		Hemarocallis citrine var. *citrina*	Flowers are citron yellow
		Hemarocallis flava	Flowers orange yellow
Hoya	Asclepiadaceae	*Hoya bella*	Flowers waxy white with a purple centre
		Hoya carnosa	Flowers pinkish white, fragrant
Oxalis	Oxalidaceae	*Oxalis hedysaroides*	Flowers bright yellow
BOP (Bird - of -Paradise)	Strelitziaceae	*Strelitzia reginae*	Sepals brilliant orange yellow, spathe boat shaped
Vanilla	Orchidaceae	*Vanila arometica*	Yellowish in bunches, fragrant

Palm species suitable for home gardening

Feather Leaved Palms :	Fan Leaved Palms :
• *Caryota mitis*	• *Livistonia chinensis*
• *Chrysalidocarpus lutescens*	• *Pritchardia pacifica*
• *Kentia belmoreana*	• *Raphis humilis*
• *Oreodoxa regia*	• *Stevensonia grandiflora*
• *Phoenix roebelenii*	• *Thrinax argentea*
	• *Washingtonia filifera*

Conifers suitable for home gardening

- Araucariaceae : *Araucaria cookii, Araucaria cunninghamii*
- Pinaceae : *Pinus wallichiana* (Bhutan pine, Himayalan pine), *Pinus sylvestris* (Gold pine)
- Taxodiaceae : *Cryptomeria japonica* (Japanese ceder)
- Pinaceae : *Cedrus deodara*
- Cupressaceae : *Cupressus macrocarpa* (Gold crest), *Juniperus chinensis* (Chinese juniper), *J. communic* (Common juniper), *Thuja orientalis, T. compacata*
- Graminae : *Bambusa glaucesceus, B. ventricosa, Arundinaria anceps* (Anceps bamboo), *A. murieliae* (Yellow bamboo), *Phyllostachys flexuosa* (Zigzag bamboo)
- Strelitiziaceae : *Ravenea madagascariensis* (Traveller's tree)

Climbers and Creepers used for home gardening

- *Monstera deliciosa*
- *Cissus antarctica*
- *Syngonium podophyllum*
- *Ficus pumila*
- *Hedera helix*
- *Clematis gouriana*
- *Antigonon leptopus*
- *Artabotrys odoratissimus*
- *Quisqualis indica*
- *Bignonia* spp.
- *Allamanda cathartica*
- *Scindapsus aureus*
- *Philodendron scandens*
- *Hoya carnosa*

Ferns suitable for home gardening

- *Dryopleris filix-mas* (Male fern)
- *Dicksonia antarctica* (Australian tree fern)
- *Polystichum aculeatum*
- *Microlepia strigosa*
- *Selaginella strigosa*
- *Polystichum setiferum*
- *Matteuccia struthiopteris*
- *Polypodium vulgare*
- *Adiantum venustum*

Cacti suitable for home gardening

- *Astrophytum myriostigma*
- *Cephalocereus senilis* – Oldman Cactus
- *Echinocactus grusonii* - Golden Barrel Cactus
- *Euphorbia millii*
- *Ferocactus latispinus*
- *Lophophora williamsii*
- *Mammillaria bocasana*
- *Opuntia microdasys*
- *Chamaecereus silvestrii*
- *Gymnocalycium mihanovichii*
- *Melocactus concinnus*
- *Notocactus graessneri*
- *Notocactus leminghausii*

Succulents suitable for home gardening

Agavaceae

- *Agave victoria – reginae*
- *Agave filifera*
- *Agave stricta*
- *Agave americana*

Liliaceae

- *Haworthia chalwinii*
- *Haworthia reinwardtii*
- *Aloe humilis*

- *Aloe aristata*
- *Aloe ferox*

Aizoaceae

- *Aptenia cordifolia*
- *Aridaria splendens*

- *Bergeranthus scapigera*

Asclepiadaceae

- *Caralluma burchardii*
- *Ceropegia juncea*

- *Duvalia elegans*
- *Echinopsis dammaniana*

Compositae

- *Kleinia acaulis*
- *Kleinia neriifolia*

- *K. tomentosa*

Bromeliaceae

- *Dyckia sulphurea*

- *Dyckia rariflora*

Crassulaceae

- *Crassula arborescens*
- *Crassula falcata*
- *Crassula perfoliata*
- *Kalanchoe beharensis*

- *Kalanchoe tomentosa*
- *Kalanchoe marmorata*
- *Cotyledon undulata*
- *Greenovia aurea*

Euphorbiaceae

- *Euphorbia bupleurifolia*
- *Euphorbia grandidens*
- *Euphorbia horrid*
- *Euphorbia millii*

- *Euphorbia polygona*
- *Euphorbia obesa*
- *Euphorbia ingens*
- *Euphorbia valida*

Portulaceae

- *Anacampseros papyracea*
- *Anacampseros ustulata*
- *Anacampseros alstonii*
- *Anacampseros rufescens*
- *Portulacaria afra*

Flowering trees suitable for home gardening

Common Name	Botanical Name	Family	Remarks
Champaka	*Michelia champaca*	Magnoliaceae	Flowers orange, yellow, strongly scented, March - Sep
Indian Medlar	*Mimusops elengi*	Sapotaceae	Dull white (or) greenish white, highly fragrant, summer months
Tree of heaven	*Amherstia nobilis*	Leguminosae	Flowers brilliant red & yellow, Jan - April
-	*Jacaranda mimosaefolia*	Bignoniaceae	Flowers bluish purple, March - April
Wild chincona	*Anthocephalus cadamba*	Rubiaceae	Flowers scented, corolla orange, calyx pale greenish
Cannon ball tree	*Couroupita guianensis*	Lecythidaceae	Flowers scented, white, yellow & pink outside & deep pink (or) crimson within
Iron wood	*Mesua ferrea*	Clusiaceae	Flowers white, fragrant
Laurel Magnolia	*Magnolia grandiflora*	Magnoliaceae	Flowers white, March – May
Bayur tree	*Pterospermum acerifolium*	Sterculiaceae	Slow growing, Flowers scented & pure white petals
Asoka tree	*Saraca asoca*	Leguminosae	Yellow and orange colour flowers
Buddhist Bauhinia	*Bauhinia variegata*	Caesalpiniaceae	White flowers, Late Jan – Mid March
Pagoda tree	*Plumeria acuminata*	Apocynaceae	White with yellow centre, scented
	Plumeria acutifolia		-
	Plumeria rubra		-
Peacock flowers	*Delonix regia*	Leguminosae	Flowers deep crimson, scarlet (or) orange
Coral jasmine	*Nyctanthes arbortristis*	Oleaceae	Scented, white flowers with orange corolla tube
Trumpet tree	*Tabebuia pallida*	Bignoniaceae	Cuban pink trumpet tree, flowers light pink
	Tabebuia rosea		Flowers rosy pink, Feb - March

Annuals suitable for home gardening

Common Name	Botanical Name	Family
Balsam	*Impatiens balsamina*	Balsaminaceae
Cosmos	*Cosmos bipinnatus*	Compositae
Blanket flower	*Gaillardia pulchella*	Compositae
Marigold	*Tagets erecta*	Asteraceae
Petunia	*Petunia hybrida*	Solanaceae
Moss rose/Table rose	*Portulaca grandiflora*	Portulaceae
Verbena	*Verbena hybrida*	Verbenaceae
Zinnia	*Zinnia elegans*	Compositae
Geranium	*Pelargonium* spp.	Geraniaceae

Bulbous ornamentals suitable for home gardening

Common Name	Botanical Name	Family
Tuberose	*Polianthes tuberosa*	Amaryllidaceae
Gerbera	*Gerbera jamesonii*	Asteraceae
Amaryllis	*Amaryllis belladona*	Amaryllidaceae
Foot ball lily	*Haemanthus* spp.	Amaryllidaceae
Indian shot	*Canna indica*	Cannaceae
Glory lily	*Gloriosa superba*	Liliaceae
Lotus	*Nelumbo lutea*	Nymphaeaceae
Water lily	*Nymphea* sp.	Nymphaeaceae
White ginger lily	*Hedychium coronarium*	Zingiberaceae

Shrubs suitable for home gardening

Common Name	Botanical Name	Family
Allamanda	*Allamanda carthatica*	Apocynaceae
Queen of night	*Cestrum nocturnum*	Solanaceae
China rose	*Hisbiscus rosa-sinensis*	Malvaceae
Mussaenda	*Mussaenda erythrophylla*	Rubiaceae
Chinese Ixora	*Ixora chienesis*	Rubiaceae

Contd...

Common Name	Botanical Name	Family
Crape Myrtle	*Lagerstroemia indica*	Lythraceae
Egyptian Star cluster	*Pentas lanceolata*	Rubiaceae
Coral plant	*Russelia juncea*	Scrophulariaceae
Crepe jasmine	*Tabernaemontana coronaria*	Apocynaceae
Jasmine	*Jasminum* spp.	Oleaceae
Cape jasmine	*Gardenia jasminoides*	Rubiaceae
Orange jasmine	*Murraya paniculata*	Rutaceae
Oleander/Nerium	*Nerium indicum*	Apocynaceae
Dwarf powderpuff	*Calliandra emarginata*	Mimosaceae

Moisture loving plants

Small plants

- *Alocasia* spp.
- *Anemone* spp.
- *Colocasia* spp.
- *Hedychium* spp.
- *Hemerocallis* spp.
- *Kniphofia* (red – hot pocker) spp.
- *Mimulus* spp.
- *Ranunculus* spp.
- *Saxifraga* spp.

Ornamental grasses & bamboos that look pretty by water side

- Pampas grass – *Cortaderia argentea*
- *Pennisetum longistylum*
- *Phyllostachys*
- Broom grass – *Thysanolaena maxima*

Plants for various landscaping component in home gardening

Plants suitable for balcony garden

- *Acalypha* spp.
- *Ananas variegata*
- *Anthurium* spp.
- *Aspidistra variegata*
- *Cryptanthus* spp.
- *Caladium* spp.
- *Dracaena* spp.
- *Gynura* spp.
- *Manihot* spp.
- *Maranta* spp.
- *Pandanus veitchi*
- *Philodendron* spp.

- *Codiaeum variegatum*
- *Coleus* spp.
- *Cordyline* spp.
- *Polyscias* spp.
- *Setcreasea pallida*
- *Xanthosoma lindenii*

Plants suitable for rock garden - Partially shaded condition

- *Phlox* spp.
- *Verbena* spp.
- *Alyssum* spp.
- *Dianthus* spp.
- *Dianthus barbatus*
- *Chlorophytum* spp.
- *Peperomia* spp.
- *Dwarf sansevieria*
- *Tradescantia* spp.
- *Zebrina* spp.
- *Fragaria* spp.
- *Ribes* spp.

Plants for rock garden – Sunny condition

- Cacti
- Succulents
- *Daedalacanthus* sp.
- *Lantana* sp.
- Miniature roses
- *Setcreasea purpurea*
- *Verbena erinoides*

Rock garden – Shade and semi-shade

- Fittonia
- Peperomia
- Selaginella
- Zebrina
- Monstera
- *Rhoeo discolor*
- *Rivinia*
- *Sansevieria*
- *Scindapsus*
- *Ruscus*
- *Tradescantia*

Plants suitable for making topiary

- *Murraya paniculata* (Kamini)
- *Lawsonia alba* (Mehandi)
- *Ixora chinensis* (Rangan)
- *Mimusops elengi* (Moulsari)
- *Duranta plumeri*
- *Juniperus chinensis*
- *Tabernaemontana coronaria* (Chandni)
- *Malphigia coccigera*
- *Casuarina equisetifolia*
- *Ficus* 'Panda'
- *Ficus benjamina* var 'Nuda'

Plants suitable for making bonsai are

- Moraceae : *Ficus benghalensis, F. elastica, F. religiosa, F. retusa, F. rumphi, F. indica, F. lyrata*

- Rutaceae : *Citrus aurantium, C. jambheri, C. reticulata, C. acida, C. maxima, Fortunella japonica, Limonia trifolia, Feronia limonia, Murraya exotica*

- Leguminosae : *Albizia lebbeck, Delonix regia, Cassia nodosa, Tamarindus indica, Bauhinia variegata*

- Sapotaceae : *Mimusops elengi, Manilhara hexandra, Achras sapota*

- Oxalidaceae : *Averrhoea carambola*

- Myrtaceae : *Psidium cattleyanum*

Plants suitable for dish (or) trough garden

- Araucaria
- Agaloenema
- *Begonia rex*
- *Calathea*
- *Coleus* spp.
- *Peperomia* spp.
- *Zebrina pendula*
- *Chlorophytum* spp.
- *Hedera helix*
- *Maranta luconneura*
- *Pilea muscosa*
- *Sansevieria* spp.
- *Scindapsus* spp.
- *Selaginella* spp.

Plants suitable for dry & hot conditions

- Chlorophytum
- *Ficus elastica*
- *Cacti*
- *Pedilanthus*
- *Aralia*
- *Peperomia*

Plants suitable for cool room

- *Cissus antactica*
- *Setcreasea purpurea*
- *Hedera helix*

Plants for small pots

- *Zebrina pendula*
- *Maranta luconeura*
- *Pilea muscosa*
- *Pilea cadierei*
- *Impatiens sultani*
- *Chlorophytum*
- Cacti and Succulents
- *Nasturtium*

Hardy plants for drought condition

- *Aspidistra* sp.
- *Dieffenbachia* sp.
- *Pedilanthus* sp.
- *Philodendron* sp.
- *Dracaena* sp.

- *Chlorophytum* sp.
- *Sansevieria* sp.
- *Pandanus* sp.
- *Monsteria* sp.
- *Ficus elastia*
- *Scindapsus* sp.

Plants for bowl & terrarium

- *Asparagus plumosus*
- Aglaonema spp.
- *Maranta luconeura*
- *Calathea illustris*
- *Begonia rex*
- *Ficus pumila*
- *Fittonia verchaffeltii*
- *Peperomia obtusifolia*

- *Philodendron* spp.
- *Pilea muscosa*
- *Asplenium* spp.
- *Tradescantia* spp.
- *Selaginella* spp.
- *Saintpaulia* spp.
- *Zebrina pendula*

Plants for hanging basket

- *Zebrina pendula*
- *Scindapsus*
- *Nasturtium*
- *Pilea muscosa*
- *Chlorophytum*
- *Asparagus*

- *Nephrolepsis exaltata*
- *Pilea cadierei*
- *Scindapsus aureus*
- *Zebrina pendula*
- *Begonia* sp.

Plants for dark corner

- Aspidistra
- Sansevieria
- Selaginella
- Monstera
- Maranta

- Tradescantia
- Scindapsus
- *Zebrina pendula*
- *Philodendron*
- *Araucaria*

Plants for east / west window

East Window	West Window
• Araucaria	• Pandanus
• Anthurium	• *Zebrina pendula*
• Begonia	• Dieffenbachia
• Caladium	• *Cissus antarctica*
• Tradescantia	• *Ficus elastica*
• Ferns	• Gloxinia
• Dracaena	• *Euphorbia milii*

Plants for South window

• Acalypha sp.	• *Coleus* sp.
• *Beloperone guttata*	• *Lilium longiflorum*
• *Cacti* sp.	• *Poinsettia* sp.
• *Bromeliads* sp.	• *Geranium*

Plants for North window

• *Aspidistra* sp.	• *Monstera* sp.
• *Aglaonema* sp.	• *Peperomia* sp.
• *Dieffenbachia* sp.	• *Philodendron* sp.
• *Hedera helix*	• *Selaginella* sp.
• *Chlorophytum* sp.	• *Sansevieria* sp.
• *Scindapsus* sp.	• *Zebrina pendula*

Plants for water garden

- Oxygenating plants : Sagittaria, *Vallisneria, Carlomba carolinana, Anacharis, Elodea*
- Water plants: Water lilies (Nymphaea), *Nymphaea capensis* (Deep blue), *N. caerulea* (Pale blue), *N. pubescens* (White), *N. rubra* (Large double, deep red), *N. stellata* (medium to large flowers, pale blue), *Nelumbium speciosum, Typha latifolia, Typha angustata*

Plants for marshy areas

• Arum lily - *Calla palustria*	• *Irises – Iris foliosa*
• *Cyperus – Cyperus alternifolius*	• *Saxifraga pellata*

Plants for group arrangement

- Selaginella
- *Syngonium podophyllum*
- *Euphorbia milii*
- *Pilea cadierei*
- *Calathea zebrina*
- Sansevieria

Plants which can be used as specimen plants

- Dieffenbachia
- Aglaonema
- Aralia
- Sansevieria
- Philodendron

Plants which can be used as large specimen plants

- *Ficus benjamina*
- *Ficus elastica*
- *Aralia elegantissima*
- *Araucaria excelsa*
- *Howea forsteriana*

Plants suitable for Window box gardening

Sunny place – Winter annuals - Aster, Phlox, Petunia, Dianthus, Pansy, Verbena and Nasturtium

Summer & rainy season annuals – Dwarf sunflower, Balsam, Zinnia, Zephyranthus

Semi-shady place– Dracaena, Dieffenbachia, Maranta, Aglaonema, Caladium, Coleus, Codiaeum, *Ficus elastica*

Shady place – Sciandapsus, Pedilanthus, Philodendran, Tradescantia

Plants for bottle garden

- *Aglaonema commutatum*
- *Dracaena sanderi*
- *Zebrina pendula*
- *Peperomia obtusifolia*
- *Cordyline terminalis*
- *Fittonia verschafelti*
- *Asplenium nidus*
- *Selaginella*

Plants for dish or bowl garden

- Cacti – Rebutia, Lobivia, Notocactus, Echinopsis, Mammillardia
- Succulents – Kalanchoe, Bryophyllum, Sedum, *Aloe variegata*

Plants for door-steps

- Aralia
- Codiaeum
- Phildendron
- Monstera
- *Ficus elastica*
- *Howea forsteriana*

Plants for entrance of a home

- Aralia
- *Calathea zebrina*
- *Chamaedorea elegans*
- *Dieffenbachia picta*
- *Codiaeum variegatum*
- *Monstera deliciosa*
- Aralia

Plants for kitchen room

- Sansevieria
- *Aspidistra lucida*

Plants for verandah

1. Reflected sunlight for 3-4 hours

- *Syngonium podophyllum*
- *Hedera helix*
- *Polypodium aureum*
- *Maranta leuconeura*
- *Dracaena sanderiana*
- Peperomia
- Alocasia
- Philodendron
- Codiaeum
- Aralia
- Aglaonema

2. Direct sunlight for 3-4 hours

- *Coleus blumei*
- *Codiaeum variegatum* var. pictum
- *Setcreasea purpurea*
- *Cordyline terminalis*
- *Tradescantia fluminensis*
- *Strobilanthes*

3. Dark and shade corner

- Sansivieria
- *Aspidistra lucida*

Plants for balcony (Area which receives direct sunlight for few hours)

- *Coleus blumei*
- *Codiaeum variegatum*
- *Ficus elastica* var. decora
- *Opuntia microdasys*

- *Pilea cadierei*
- *Impatiens sultani*
- *Gloxinia*
- *Kalanchoe*

Plants for side walls

- Chlorophytum
- Saintpaulia
- Nasturtium

- Petunia
- *Pilea cadieri*

Plants for dining table

- Small sized pot plants – *Fittonia verschafeltii, Asplenium nidus,* Scindapsus
- Glass container with clean water - Scindapsus

Plants for hall arrangement

Dark corner

- *Aspidistra lucida*
- *Sansevieria trifasciata* var. laurentii

- *Anthurium magnificus*
- Aglaonema

Shady areas

- *Cissus antartica*
- *Ficus benjamina*

- *Dieffenbachia picta*
- Feathered leaved palm

Windowsill

- Saintpaulia
- Poinsettia
- Coleus

- Cacti
- Bonsai

Plants for windows which receive indirect (or) reflected sunlight

- Syngonium
- Dieffenbachia
- Aglaonema
- Calathea

- Monstera
- Maranta
- Sansevieria
- Philodendron

Plants for windows which receives direct sunlight for few hours

- *Scindapsus aureus*
- Impatiens
- Kalanchoe

Plants for drawing room

- *Philodendron scandens*
- *Maranta bucolara*
- *Anthurium magnificus*
- *Caladium bicolor*

Plants for dark corner

- Sansevieria
- Aglaonema
- *Victoria dracaena*

Group plants for corner

- *Calathea zebrina*
- *Syngonium podophyllum*
- *Pilea cadierei*

Plants for reading room

- Juniperus
- Asparagus
- Sansevieria as stony plant

3

Factors to be Considered for Growth of House Plants

Factors Affecting Plant Growth

Plant growth is affected by light, temperature, humidity, water, nutrition, and soil.

Light

Light is needed for plants to produce food and survive. In general, if more light is available in a particular location, the more will be the food production which helps the plant to grow faster. Different plants require different light intensities. Intensity (or quality) of light is difficult to measure without a light meter. It is usually measured in units of foot candles or lux. The two important factors for providing light to a house plant are intensity and duration. Using the light readings, indoors can be divided into four areas, which have the following light levels for 8 hours per day:

1. Sunny light areas: at least 4 hours of direct sunlight

2. High-light areas: over 200 ft-c (2,150 Lux) but not direct sunlight

3. Medium-light areas: 75 ft-c – 200 ft-c (800-2,150 Lux)

4. Low-light areas: 25 ft-c – 75 ft-c (270-800 Lux)

The duration of light exposure is as important as the intensity. Quality exposure of 8 to 16 hours is ideal for most plants. Windows are the most common sources of light for houseplants. In the Northern Hemisphere, south-facing windows have the most sun exposure, while western, eastern, and north-facing windows have progressively less exposure. Natural sunlight through windows is affected by seasonal changes, cloud cover, and window treatments. The length of time that light is provided will determine how the plant grows. Providing 16 hours of light/day will promote strong roots, stems and abundant leaves. Decreasing that amount to 12 hours of light/day will signal that the short days of winter months, for that plant will focus more on flower production and less on vegetative growth.

Photoperiodism must also be considered, since some plants such as **Poinsettia** and **Christmas cactus** are influenced by either decreasing or increasing daylight hours. Outdoors, the light levels on a bright day range from 10,000 ft-c (1,07,500 Lux) in an open sunny area to 250 ft-c (2700 Lux) or less in the shade of a large tree. Artificial light sources can provide an alternative or supplement to window lighting. Fluorescent lighting provides excellent light quality whereas standard incandescent bulbs do little to promote plant growth. Though incandescent bulbs are very cheap, there is no use for the plants, but the cost of the electricity will outweigh the cost of purchasing multiple incandescent bulbs and fixtures in the long run. LED or Light Emitting Diode lights which produce the photosynthetic optimum red (640-670 nm) and blue light (430-460 nm), they have long life expectancies, and efficiency, 100 watts producing 3,400 lumens output from a unit 400x212x62mm, they are also rather cool, so you can have them quite close to the plants. Fluorescents are mass marketed. CFL fluorescents are the cheapest option, but more than a couple bulbs are almost always required to be running at once. "Cool", or "blue", fluorescent lights at 6500k provide the light needed for lush green foliage plants, while "warm", or "red", fluorescent lights at 3000k provide the light needed for blooming flowers and fruit production. Warm whites are better for flowering plants while cool whites are more suitable for green, leafy growth. When used together, these bulbs are closer to the full spectrum light that comes from the sun, although less powerful.

Windows with eastern exposure within the home generally provide the best light and temperature conditions for most indoor plant growth because plants receive direct morning light from sunrise until nearly midday. Foot-candle readings at these windows can reach 5,000-8,000. As the morning progresses, the direct sun recedes from the room. An eastern room is cooler than southern or western rooms because the house absorbs less radiant heat. Light from the east is cooler than that from the south or the west, and thus it causes less water loss from the plants. Windows with southern exposure give the largest variation of light and temperature conditions.

Windows with northern exposure provide the least light and the lowest temperature. Because out of the four exposures, the northern exposure receives the least light and heat year round. Because of the low-light levels, maintaining healthy plants can be a challenge. A northern windowsill can measure light levels as low as 200 ft-c on a clear winter day, which is optimal for some plants, such as the African violet. This exposure is best for plants with green foliage because the coloration on variegated foliage tends to disappear under low-light conditions. Although most plants grown indoors will not grow in a northern room, they may tolerate it for short periods of time. Seasons change the amount of natural light entering through windows.

In the summer, when the sun is farther north than it is in the winter, the sun rises at a sharp angle in the morning and is high in the sky by noon. Therefore, sunlight penetrates farther into a room during winter. Direct light comes into a south window only at midday. If there is a wide overhang covering the windows outside, the sun may not enter the room at all. The sun at noon on a summer day may measure 10,000 ft-c. Indoors, however, a southern window with wide eaves on the outside will receive about the same amount of light as a window with northern exposure. Southern and western exposures are interchangeable for most plants. In the winter, most plants, except those with definite preference for northern exposure, can be placed in a room with southern exposure.

Most flowering plants, and some sun-loving foliage plants, need to be within 3 feet of a sunny, south-facing window. Plants that prefer bright, indirect light can be located 3 to 5 feet away from a south-facing window, or within 3 feet of an east- or west-facing window. Plants that thrive in diffused light can be placed 6 to 8 feet away from a south-facing window, or within a foot of a north-facing window. In that location they'll receive about 25 percent of the light they would get if they were in front of the sunny, south-facing window. During the winter months, all the plants will be moved closer to the window in order to compensate for the decrease in light. Most plants perform best when they receive 12 to 16 hours of light per day. If you want to keep your plants blooming during the short days of winter, you may need to provide supplemental lighting.

Many foliage plants are native to tropical rain forests, where light levels are low. These plants can be easily injured if exposed to strong light. Symptoms of over-exposure are upright leaves and bleached, scorched leaves. Do not place high-light sensitive plants in direct sunlight (on a porch or in front of a window). In this example, Chinese Evergreen (*Aglaonema*) and Dumb Cane (*Dieffenbachia*) show symptoms of high-light damage. While adequate light is crucial for plant growth, too much light can be damaging.

1. Plants suitable for sunny light areas (at least 4 hours direct sun) are,

Botanical Name	Common Name
Abutilon hybridum	Flowering maple
Acalypha hispida	Acalypha
Agave americana 'Marginata'	Variegated century plant
Agave victoriae-reginae	Queen agave
Allamanda cathartica	Allamanda
Aloe arborescens	Aloe
Aloe barbadensis	Aloe
Aloe brevifolia	Aloe
Beaucarnea recurvata	Pony tail palm
Begonia semperflorens	Wax begonia
Bougainvillea spp.	Bougainvillea
Cereus peruvianus	Peruvian apple cactus
Chrysanthemum morifolium	Chrysanthemum
Codiaeum variegatum	Croton
Crassula falcata	Propeller plant
Crassula hemispherica	Arab's turban
Crassula lycopodioides	Toy cypress
Dyckia brevifolia	Miniature agave
Dyckia fosterana	Silver and gold dychia
Echeveria agavoides	Molded wax
Echeveria elegans	Mexican snowball
Echinocereus reichenbachii	Lace cactus
Euphorbia mammillaris	Corncob cactus
Euphorbia milii splendens	Crown of thorns

Contd...

Botanical Name	Common Name
Euphorbia pulcherrima	Poinsettia
Euphorbia tirucalli	Milk bush
Ficus benjamina	Weeping fig
Ficus elastica 'Decora'	Rubber plant
Ficus lyrata	Fiddle – leaf fig
Haworthia cuspidata	Star window plant
Haworthia fosciata	Zebra haworthia
Haworthia truncata	Clipped window plant
Hibiscus rosa-sinensis	Chinese hibiscus
Ixora coccinea	Ixora
Jatropha integerrima	Peregriam
Justicia brandegeana (syn : Beloperone guttata)	Flamingo plant, Kings crown, Plume plant
Kalanchoe blossfeldiana	Christmas kalanchoe
Kalanchoe tomentosa	Panda plant
Malvaviscus arboreus	Turk's cap
Mammillaria bocasana	Powder puff
Opuntia vilis	Little tree cactus
Opuntia vulgaris	Irish mittens
Pelargonium hortorum	Horse geranium
Pelargonium pelltatum	Ivy geranium
Pentas lanceolata	Egyptian star cluster
Ruellia graeciznas	Red – spray ruellia
Sedum spectabile	Showy sedum
Sempervium arachniodeum	Cow web house leek
Setcreasea pallida	Purple heart

2. Plants suitable for high – light areas (over 200 foot candles (2150 Lux) – not direct sun)

Botanical Name	Common Name
Acorus calamus	Sweet flag
Acorous gramineus	Miniature sweet flag
Adiantum raddianum	Maiden hair fern
Aechme miniata 'Discolor'	Purplish coral berry
Aechmea fasciata	Silver vase
Aechmea 'Royal wine'	Royal wine bromeliad
Anthurium clarinervium	Dwarf crystal anthurium
Anthurium hookeri	Bird's nest anthurium
Anthurium scherzerianum	Flamingo flower
Aphelandra squarrosa	Zebra plant
Araucaria heterophylla	Norfolk island pine
Ardissa crenata	Ardisia
Asparagus densiflorus 'Myers'	Plume asparagus
Asparagus densiflorus 'Sprengeri'	Fax tail fern
Asparagus falcatus	Sickle thorn
Astrophytum myriostigma	Bishop's cap
Begonia cubensis	Cuban holly
Begonia metallica	Metallic leaf ban begonia
Begonia rex-cultorum	Rex begonia
Begonia semperflorens	Wax begonia
Billbergia nutans	Queens tears
Billbergia pyramidalis	Urn plant
Billbergia zebrina	Zebra plant
Brassaia actinophylla	Schefflera
Brassaia arboricola	Dwarf schefflera
Caladium spp.	Caladium
Calathea insignis	Rattlesnake plant
Calathea makoyana	Peacock plant

Contd...

Botanical Name	Common Name
Calathea roseopicta	Rose calathea
Calceolaria crenatiflora	Slipper wort
Caryota mitis	Fish tail palm
Catharanthus roseus	Madagascar periwinkle
Ceropegia woodii	Rosary vine
Chamaerops humilis	European fan palm
Chlorophytum comosum 'Variegatum'	Variegated spider plant
Chlorophytum comosum 'Vittatum'	Spider plant
Chrysalidocarpus lutescens	Areca palm
Cissus antartica	Kangaroo vine
Cissus rhombifolia	Grape leaf ivy
Cissus rotundifolia	Wax cissus
Cissus striata	Miniature grape ivy
Clivia miniata 'Grandiflora'	Kafir lily
Coleus blumei	Coleus
Colummea hybrida	Gold fish plant
Cordyline terminalis	Ti plant
Crassula argentea	Jade plant
Crassula falcata	Propeller plant
Crassula hemisphaerica	Arab'e turban
Crassula lycopodioides	Toy cypress
Crassula schmidtii	Red flowering crassula
Crossandra infundibuliformis	Crossandra
Cryptanthus fosteranus	Stiff pheasant leaf
Cryptanthus bivittatus 'Minor'	Dwarf rose stripe star
Cryptanthus zonatus	Zebra plant
Cyrtominum falcatum 'Rochfordianum'	House holly fern
Davallia feieensis	Rabbit's foot fern
Dieffenbachia 'Exotica Perfection'	Exotica perfection
Dizygotheca elegantissima	False aralia

Contd...

Botanical Name	Common Name
Dracaena deremensis 'Janet Craig'	Janet craig
Dracaena deremensis 'Warneckii'	Warneckii
Dracaena fragrans 'Massangeana'	Corn plant
Dracaena marginata	Marginata
Dracaena surculosa	Gold dust dracaena
Dyckia brevifolia	Miniature agave
Dyckia fosterana	Silver and gold dyckia
Echeveria agavoides	Molded wax
Echeveria elegans	Mexican snowball
Echinocereus reichenbachii	Lace cactus
Epidendrum atropurpureum	Spice orchid
Epiphyllum hybrida	Golden pothos
Epipremnum aureum	Golden pothos
Epipremnum aureum 'Marble Queen'	Marble queen
Episcia cupreata	Flame violet
Episcia dianthiflora	Lace – flower vine
Episcia reptans	Scarlet violet
Euphorbia coeralescens	Blue euphorbia
Euphorbia pulcherrima	Poinsettia
Euphorbia tirucalli	Milk bush
Ficus benjamina	Weeping fig
Ficus deltoidea	Mistletoe fig
Ficus elastica 'Decora'	Rubber plant
Ficus lyrata	Fiddle leaf fig
Ficus pumila 'Minima'	Dwarf creeping fig
Ficus retusa	Cuban laurel
Ficus sagittata	Rooting fig
Ficus willdemaniance	Dwarf fiddle leaf fig
Fittonia verschaffeltii	Red nerved fittonia
Fittonia verschaffeltii argyroneura	Silver nerved fittonia

Contd...

Botanical Name	Common Name
Fuchsia hybrida	Fuchsias
Gasteria hybrida	Ox tongue
Guzmania ligulata 'Major'	Scarlet star
Guzmania monostachia	Striped torch
Gynura aurantiaca 'Purple Passion'	Purple passion
Haworthia cuspidata	Zebra haworthia
Haworthia subfasciata	Little zebra plant
Haworthia truncata	Clipped window plant
Hedera helix	English ivy
Hedera canariensis	Algerian ivy
Hemigraphis alternata	Waffle plant
Hippeastrum hybrida	Amaryllis
Howea belmoreana	Belmore sentry palm
Howea forsterana	Kentia palm
Hoya carnosa 'Variegata'	Wax plant
Hoya kerrii	Sweetheart hoya
Hyacinthus orientalis	Hyacinth
Impatiens wallerana 'Variegata'	Busy lizzie impatiens
Justicia brandegeana	Shrimp plant, flamingo plant, plume plant
Kalanchoe blossfeldiana	Christmas kalanchoe
Kalanchoe pumila	Dwarf purple kalanchoe
Kalanchoe tomentosa	Panda plant
Mammillaria bocasana	Powder puff
Maranta leuconeura erythroneura	Red nerve plant
Maranta leuconeura kerchovina	Prayer plant
Monstera deliciosa	Philodendron pertusum
Nautilocalyx lynchii	Black alloplectus
Neoregelia carolinae 'Tricolor'	Tricolor bromeliades
Neoregelia spectabilis	Finger nail plant
Neoregelia zonata	Zonata

Contd...

Botanical Name	Common Name
Nephrolepis exaltata 'Bostoniensis'	Boston fern
Nephrolepis exaltata 'Fluffy Ruffles'	Fluffy ruffles
Nidularium innocentii nana	Miniature bird's nest
Opuntia vilis	Little tree cactus
Opuntia vulgaris	Irish mittens
Oxalis flava	Finger oxalis
Oxalis rubra	Red oxalis
Pachystachys clutea	Yellow shrimp plant
Paphiopedilum hybrida	Lady slipper orchid
Pedilanthus tithymaloides 'Variegatus'	Devil's backbone
Pelargonium hortorum	House geranium
Pelargonium peltatum	Ivy geranium
Pellaea rotundifolia	Button fern
Pellionia pulchra	Satin pellionia
Peperomia crassifolia	Leather peperomia
Peperomia caperata	Emerald ripple
Peperomia obtusifolia	Baby rubber tree
Philodendron 'Emerald Queen'	Emerald queen
Philodendron 'Florida'	Florida
Philodendron scandens oxycardium	Heart – leaf philodendran
Philodendron selloum	Selloum
Phoenix roebelenii	Pigmy date palm
Pilea cadierei	Aluminium plant
Pilea microphylla	Artillery plant
Platycerium bifurcatum	Staghorn fern
Plectranthus australis	Swedish ivy
Podocarpus macrophyllus	Podocarpus
Polyscias balfouriana 'Marginata'	Variegated balfour aralia
Polyscias fruticosa	Lady palm

Contd...

Botanical Name	Common Name
Rhododendron hybrida	Azalea
Ruellia graeciznas	Red – spray ruellia
Saintpaulia hybrida	African violet
Sainpaulia parva	Parva sansevieria
Sainpaulia trifasciata 'Hahnii'	Bird's nest sanseivieria
Sainpaulia trifasciata 'Laurentii'	Gold banded sansevieria
Saxifraga stolonifera	Strawberry geranium
Schlumbergera bridgesii	Christmas cactus
Schlumbergera truncata	Christmas cactus
Sedum spectabile	Showy sedum
Sempervivum arachniodeum	Cow web house leek
Setcreasea pallid 'Purple Heart'	Purple heart
Sinningia speciosa	Gloxinia
Soleirolia soleirolii	Baby tears
Spathiphyllum 'Clevelandii'	Peace lily
Spathiphyllum 'Mauna Loa'	Mauna loa
Stapelia nobilis	Carrion flower
Streptocarpus hybridus	Cape primrose
Strobilanthes dyeranum	Persian shield
Syngonium podophyllum	Nephthytis
Trillandsia bulbosa	Dancing bulb
Trillandsia lindenii	Blue – flowered torch
Tradescantia blossfeldiana	Flowering inch plant
Tradescantia sillamontana	White velvet plant
Vriesea splendens	Flaming sword
Yucca elephantipes	Spineless yucca
Zebrina pendula	Wondering jew

3. Plants which are suitable for medium light areas (75 foot candles (807 Lux) to 200 foot candles(2153 Lux))

Botanical Name	Common Name
Acorus calamus	Sweet flag
Acorous gramineus	Miniature sweet flag
Adiantum raddianum	Maiden hair fern
Aechmea fasciata	Siver vase
Aechmea miniata 'Discolor'	Purplish coral berry
Aechmea 'Royal wine'	Royal wine bromeliad
Aglaonema modestum	Chinese evergreen
Aglaonema modestum 'Silver King'	Silver king
Anthurium clarinervium	Dwarf crystal anthurium
Anthurium hookeri	Bird's nest anthurium
Anthurium scherzerianum	Flamingo flower
Araucaria heterophylla	Norfolk island pine
Ardissa crenata	Ardisia
Asparagus densiflorus 'Myers'	Plume asparaguas
Asparagus densiflorus 'Sprengeri'	Fox tail fern
Asparagus falcatus	Sickle thorn
Asplenium daucifolium	Mother fern
Asplenium nidus	Bird's nest fern
Begonia cubensis	Cuban holly
Begonia rex-cultorum	Rex begonia
Billbergia zebrina	Zebra plant
Billbergia nutans	Queen's tears
Billbergia pyramidalis	Urn plant
Brassaia actinophylla	Schefflera plant
Brassaia arboricola	Dwarf schefflera plant
Calathea insignis	Rattlesnake plant

Contd...

Botanical Name	Common Name
Calathea makoyana	Peacock plant
Calathea roseopicta	Rose calathea
Caryota mitis	Fishtail palm
Ceropegia woodii	Rosary vine
Chamaerops humilis	European fan palm
Chlorophytum comosum 'Variegatum'	Variegated spider plant
Chlorophytum comosum 'Vittatum'	Spider plant
Chrysalidocarpus lutescens	Areca palm
Cissus antartica	Kangaroo vine
Cissus rhombifolia	Grape leaf ivy
Cissus striata	Miniature grape ivy
Coleus blumei	Coleus
Colummea hybrida	Gold fish plant
Crassula schmidtii	Red flowering crassula
Crassula teres	Rattlesnake tail
Cyrtominum falcatum 'Rochfordianum'	House holly fern
Davallia feieensis	Rabbit's foot fern
Dieffenbachia 'Exotica Perfection'	Exotica perfection
Dieffenbachia maculata	Spotted dumb cane
Dizygotheca elegantissima	False aralia
Dracaena fragrans 'Massangeana'	Corn plant
Epipremnum aureum	Golden pothos
Epipremnum aureum 'Marble Queen'	Marble queen
Euphorbia coeralescens	Blue euphorbia
Fatshedera lizei	Botanical wonder plant
Ficus benjamina	Weeping fig
Ficus deltoidea	Mistletoe fig
Ficus pumila 'Minima'	Dwarf creeping fig

Contd...

Botanical Name	Common Name
Ficus retusa	Cuban laurel
Ficus sagittata	Rooting fig
Ficus willdemaniance	Dwarf fiddle –leaf fig
Fittonia verschaffeltii	Red nerved fittonia
Fittonia verschaffeltii argyroneura	Silver nerved fittonia
Gynura aurantiaca 'Purple Passion'	Purple passion
Hedera helix	English ivy
Hemigraphis alternata	Waffle plant
Howea belmoreana	Belmore sentry palm
Howea forsterana	Kentia palm
Hoya carnosa 'Variegata'	Wax plant
Impatiens wallerana 'Variegata'	Busy lizzie impatiens
Maranta leuconeura erythroneura	Red nerve plant
Maranta leuconeura kerchovina	Prayer plant
Monstera deliciosa	Philodendron pertusum
Neoregelia carolinae 'Tricolor'	Tricolor bromeliad
Neoregelia spectabilis	Finger nail plant
Neoregelia zonata	Zonata
Nephrolepis exaltata 'Bostoniensis'	Boston fern
Nephrolepis exaltata 'Fluffy Ruffles'	Fluffy ruffles
Pachystachys clutea	Yellow shrimp plant
Paphiopedilum hybrida	Lady slipper orchid
Pellaea rotundifolia	Devils backbone
Pellionia pulchra	Satin pellionia
Peperomia caperata	Emerald ripple
Peperomia crassifolia	Leather peperomia
Peperomia obtusifolia	Baby rubber tree

Contd...

Botanical Name	Common Name
Philodendron bipennifolium	Fiddle – leaf philodendron
Philodendron 'Emerald Queen'	Emerald queen
Philodendron 'Florida'	Florida
Philodendron scandens oxycardium	Heart leaf philodendron
Philodendron selloum	Selloum
Phoenix roebelenii	Pigmy date palm
Pilea cadierei	Aluminum plant
Pilea microphylla	Artillery plant
Platycerium bifurcatum	Staghorn fern
Plectranthus australis	Swedish ivy
Podocarpus macrophyllus	Podocarpus
Polyscias balfouriana 'Marginata'	Variegated balfour aralia
Polyscias fruticosa	Lady palm
Saintpaulia hybrida	African violet
Sainpaulia parva	Parva sansevieria
Sainpaulia trifasciata 'Hahnii'	Bird's nest sansevieria
Sainpaulia trifasciata 'Laurentii'	Gold-banded sansevieria
Saxifraga stolonifera	Strawberry geranium
Schlumbergera bridgesii	Christmas cactus
Sinningia speciosa	Gloxinia
Spathiphyllum 'Clevelandii'	Peace lily
Spathiphyllum 'Mauna Loa'	Mauna loa
Syngonium podophyllum	Nephthytis
Tradescantia blossfeldiana	Flowering inch plant
Tradescantia sillamontana	White velvet
Zebrina pendula	Wondering jew

4. Plants suitable for low light areas (25 foot candles (269 Lux) to 75 foot candles (807 Lux))

Botanical Name	Common Name
Aglaonema modestum	Chienese evergreen
Aglaonema modestum 'Silver King'	Silver king
Dracaena deremensis 'Janet Craig'	Janet craig
Dracaena deremensis 'Warneckii'	Warneckii
Howea forsterana	Kentia palm
Monstera deliciosa	Philodendron pertusum
Philodendron bipennifolium	Fiddle – leaf philodendron
Philodendron 'Emerald Queen'	Emerald queen
Philodendron 'Florida'	Florida
Philodendron scandens oxycardium	Heart leaf philodendron
Philodendron selloum	Selloum
Sainpaulia trifasciata 'Hahnii'	Birds nest sansevieria
Sainpaulia trifasciata 'Laurentii'	Gold banded sansevieria
Syngonium podophyllum	Nephthytis

Temperature

In their natural habitat, most plants experience a day-to-night temperature fluctuation of at least 10°F. But in indoors, they will benefit from having a similar temperature differential. Most plants also expect a resting period each year; in fact, some flowering plants actually require a period of dormancy before they will set bud and flower.

To simulate this resting period, watering and fertilizer application could be withheld during winter season, when the intensity and duration of natural light is lowest. Once the day length begins to increase, start the process of watering and fertilizer application. The temperature required for the growth of plants varies as follows.

1. Cool: 65°F (25°C)day ,50°F (90°C) night temperatures
2. Average: 75°F (28.5°C) day, 65°F (25°C) night temperatures
3. Warm: 85°F (32°C) day, 70°F (26.5°C) night temperatures

House plants suitable for the different temperature requirements are furnished below.

Plants which fit the category 1 (65ºF (18.33ºC) day temperature, 50ºF (10 ºC)night)

Botanical Name	Common Name
Abutilon hybridum	Flowering maple
Fatshedera lizei	Botanical wonder plant
Hedera helix	English ivy
Hedera canariensis	Algerian ivy
Hibiscus rosa-sinensis	Chinese hibiscus
Hyacinthus orientalis	Hyacinth
Rhododendron hybrida	Azalea
Sedum spectabile	Showy sedum
Sempervium arachnoideum	Cow web house leek
Begonia semperflorens	Wax begonia
Calceolaria crenatiflora	Slipper wort
Cordyline terminalis	Ti plant
Pelargonium hortorum	House geranium
Pelargonium peltatum	Ivy geranium
Saxifraga stolonifera	Strawberry geranium

Plants which suit the category 2 (Surface of media should dry before rewatering) are (75ºF (23.88 ºC) day temperature, 65ºF (18.33 ºC) night)

Botanical Name	Common Name
Acalypha hispida	Che nile plant
Achimene hybrida	Magic flower
Acorus calamus	Sweet flag
Acorous gramineus	Miniature sweet flag
Adiantum raddianum	Maiden hair fern
Aechmea miniata 'Discolor'	Purplish coral berry
Aechmea fasciata	Silver vase
Aechmea 'Royal wine'	Royal wine bromeliad

Contd...

Botanical Name	Common Name
Aglaonema modestum	Chinese evergreen
Aglaonema modestum 'Silver King'	Silver king
Anthurium clarinervium	Dwarf crystal anthurium
Anthurium hookeri	Bird's nest anthurium
Anthurium scherzerianum	Flamingo flower
Araucaria heterophylla	Norfolk island pine
Ardissa crenata	Ardisia
Asparagus densiflorus ' Myers'	Plume asparaguas
Asparagus densiflorus ' Sprengeri'	Fox tail fern
Asparagus falcatus	Sickle thorn
Asplenium daucifolium	Mother fern
Asplenium nidus	Bird's nest fern
Begonia cubensis	Cuban holly
Begonia rex-cultorum	Rex begonia
Billbergia zebrina	Zebra plant
Billbergia nutans	Queen's tears
Billbergia pyramidalis	Urn plant
Brassaia actinophylla	Schefflera
Brassaia arboricola	Dwarf schefflera
*Caladium*spp.	Caladium
Calathea insignis	Rattlesnake plant
Calathea makoyana	Peacock plant
Calathea roseopicta	Rose calathea
Calceolaria crenatiflora	Slipper wort
Carissa grandiflora 'Bonsai'	Bonsai natal plum
Carissa grandiflora 'Boxwood Beauty'	Boxwood beauty
Caryota mitis	Fish tail palm
Catharanthus roseus	Madagascar periwinkle
Ceropegia woodii	Rosary vine

Contd...

Botanical Name	Common Name
Chamaerops humilis	European fan palm
Chirita lavandulacea	Hindustan gentian
Chlorophytum comosum 'Variegatum'	Variegated spider plant
Chlorophytum comosum 'Vittatum'	Spider plant
Chrysalidocarpus lutescens	Areca palm
Cissus antartica	Kangaroo vine
Cissus rhombifolia	Grape leaf ivy
Cissus rotundifolia	Wax cissus
Cissus striata	Miniature grape ivy
Cissus rotundifolia	Wax cissus
Cissus striata	Miniature grape ivy
Clivia miniata 'Grandiflora'	Kafir lily
Coleus blumei	Coleus
Colummea hybrida	Gold fish plant
Cordyline terminalis	Ti plant
Crassula argentea	Jade plant
Crassula falcata	Propeller plant
Crassula hemisphaerica	Arab's turban
Crassula lycopodioides	Toy cypress
Crassula schmidtii	Red flowering crassula
Crassula teres	Rattle snake tail
Crossandra infundibuliformis	Crossandra
Cryptanthus fosteranus	Stiff pheasant leaf
Cryptanthus bivittatus 'Minor'	Dwarf rose stripe star
Cryptanthus zonatus	Zebra plant
Cyrtominum falcatum 'Rochfordianum'	House holly fern
Davallia feieensis	Rabbit's foot fern
Dieffenbachia 'Exotica Perfection'	Exotica perfection
Dizygotheca elegantissima	False aralia

Contd...

Botanical Name	Common Name
Dracaena deremensis 'Janet Craig'	Janet craig
Dracaena deremensis 'Warneckii'	Warneckii
Dracaena fragrans 'Massangeana'	Corn plant
Dracaena marginata	Marginata
Dracaena surculosa	Gold dust dracaena
Echeveria agavoides	Molded wax
Echeveria elegans	Mexican snowball
Echinocereus reichenbachii	Lace cactus
Epidendrum atropurpureum	Spice orchid
Epiphyllum hybrida	Golden pothos
Epipremnum aureum	Golden pothos
Epipremnum aureum 'Marble Queen'	Marble queen
Episcia cupreata	Flame violet
Episcia dianthiflora	Lace – flower vine
Episcia reptans	Scarlet violet
Euphorbia coeralescens	Blue euphorbia
Euphorbia pulcherrima	Poinsettia
Euphorbia tirucalli	Milk bush
Fatshedera lizei	Botanical wonder plant
Ficus benjamina	Weeping fig
Ficus deltoidea	Mistletoe fig
Ficus elastica 'Decora'	Rubber plant
Ficus lyrata	Fiddle leaf fig
Ficus pumila 'Minima'	Dwarf creeping fig
Ficus retusa	Cuban laurel
Ficus sagittata	Rooting fig
Ficus willdemaniance	Dwarf fiddle leaf fig
Fittonia verschaffeltii	Red nerved fittonia
Fittonia verschaffeltii argyroneura	Silver nerved fittonia

Contd...

Botanical Name	Common Name
Fuchsia hybrida	Fuchsias
Gasteria hybrida	Ox tongue
Guzmania ligulata 'Major'	Scarlet star
Guzmania monostachia	Striped torch
Gynura aurantiaca 'Purple Passion'	Purple passion
Haworthia cuspidata	Zebra haworthia
Haworthia subfasciata	Little zebra plant
Haworthia truncata	Clipped window plant
Hedera helix	English ivy
Hedera canariensis	Algerian ivy
Hemigraphis alternata	Waffle plant
Hippeastrum hybrida	Amaryllis
Howea belmoreana	Belmore sentry palm
Howea forsterana	Kentia palm
Hoya carnosa 'Variegata'	Wax plant
Hoya kerrii	Sweetheart hoya
Hyacinthus orientalis	Hyacinth
Impatiens wallerana 'Variegata'	Busy lizzie impatiens
Justicia brandegeana	Shrimp plant
Kalanchoe blossfeldiana	Christmas kalanchoe
Kalanchoe pumila	Dwarf purple kalanchoe
Kalanchoe tomentosa	Panda plant
Mammillaria bocasana	Powder puff
Manettia inflata	Firecracker plant
Maranta leuconeura erythroneura	Red nerve plant
Maranta leuconeura kerchovina	Prayer plant
Neoregelia carolinae 'Tricolor'	Tricolor bromeliad
Neoregelia spectabilis	Finger nail plant
Neoregelia zonata	Zonata

Contd...

Botanical Name	Common Name
Nephrolepis exaltata ' Bostoniensis'	Boston fern
Nephrolepis exaltata 'Fluffy Ruffles'	Fluffy ruffles
Opuntia vilis	Little tree cactus
Opuntia vulgaris	Irish mittens
Oxalis flava	Finger oxalis
Oxalis rubra	Red oxalis
Pachyphytum oviferum	Pearly moonstones
Pachystachys clutea	Yellow shrimp plant
Paphiopedilum hybrida	Lady slipper orchid
Pedilanthus tithymaloides 'Variegatus'	Devil's backbone
Pelargonium hortorum	House geranium
Pelargonium peltatum	Ivy geranium
Pellaea rotundifolia	Button fern
Pellionia pulchra	Satin pellionia
Peperomia caperata	Emerald ripple
Peperomia crassifolia	Leather peperomia
Peperomia obtusifolia	Baby rubber tree
Philodendron bipennifolium	Fiddle – leaf philodendron
Philodendron 'Emerald Queen'	Emerald queen
Philodendron 'Florida'	Florida
Philodendron scandens oxycardium	Heart leaf philodendron
Philodendron selloum	Selloum
Phoenix roebelenii	Pigmy date palm
Pilea cadierei	Aluminum plant
Pilea microphylla	Artillery plant
Platycerium bifurcatum	Staghorn fern
Plectranthus australis	Swedish ivy
Podocarpus macrophyllus	Podocarpus

Contd...

Botanical Name	Common Name
Polyscias balfouriana 'Marginata'	Variegated balfour aralia
Polyscias fruticosa	Lady palm
Rhododendron hybrida	Azalea
Ruellia graeciznas	Red – spray ruellia
Saintpaulia hybrida	African violet
Sainpaulia parva	Parva sansevieria
Sainpaulia trifasciata 'Hahnii'	Bird's nest sanseivieria
Sainpaulia trifasciata 'Laurentii'	Gold banded sansevieria
Saxifraga stolonifera	Strawberry geranium
Sedum spectabile	Showy sedum
Sempervivum arachniodeum	Cow web house leek
Setcreasea pallid 'Purple Heart'	Purple heart
Sinningia speciosa	Gloxinia
Soleirolia soleirolii	Baby tears
Spathiphyllum 'Clevelandii'	Peace lily
Spathiphyllum 'Mauna Loa'	Mauna loa
Stapelia nobilis	Carrion flower
Streptocarpus hybridus	Cape primrose
Strobilanthes dyeranum	Persian shield
Syngonium podophyllum	Nephthytis
Trillandsia bulbosa	Dancing bulb
Trillandsia lindenii	Blue – flowered torch
Tolmiea menziesii	Piggyback plant
Tradescantia blossfeldiana	Flowering inch plant
Tradescantia sillamontana	White velvet plant
Vriesea splendens	Flaming sword
Yucca elephantipes	Spineless yucca
Zebrina pendula	Wondering jew

Plants which fit the category 3 is (warm: 85ºF (17 ºC) day temperature, 70ºF (14 ºC) night)

Botanical Name	Common Name
Carissa grandiflora 'Bonsai'	Bonsai natal plum
Carissa grandiflora 'Boxwood Beauty'	Boxwood beauty
Episcia cupreata	Flame violet
Episcia dianthiflora	Lace – flower vine
Episcia reptans	Scarlet violet
Ficus elastic 'Decora'	Rubber plant

Relative Humidity

All rooms in the house can differ massively in their humidity levels, bathrooms and kitchens are generally the most humid rooms while living rooms and bedrooms being drier. Hall is the driest part of the home. Most plants grow well when the relative humidity is 50 percent or higher, though they can usually survive at 30 to 40 percent. If the air is much drier than that, they are unable to absorb enough water through their roots to keep up with the water lost through their leaves

During summer, plants are suffering from dry air, they will show the symptoms of brown leaf tips, yellow edged wilting leaves, leaf drop and flowers shrivelling. In such a case, increasing humidity is the only option to save the house plants especially during summers. For which all the interior plants are placed together or place the pots in a group. This will increase the humidity around the group as moisture is evaporated from the compost and transpired from the leaves. Another method is placing the plants on a gravel-filled tray that contains about 1/4 inch of water. As the water evaporates, it will humidify the air around your plants. But the pots don't be placed directly in the water. Other best option is placing plants like Sansieveria, Beaucarnea, *Ficus elastica* and most palms which perform well under such dry environment.

During winter, humidity level will be 10 to 20 percent in indoors. So, grouping is not required, if we place like that, it favours the growth of fungus due to lack of ventilation. Plants will come under any of the following three categories. They are

1. High: 50% or higher

2. Average: 25% to 49%

3. Low: 5% to 24%

Plants which are classified under the category 1 (High: 50% or higher)

Botanical Name	Common Name
Adiantum raddianum	Maiden hair fern
Alloplectus nummularia	Miniature pouch flower
Anthurium clarinervium	Dwarf crystal anthurium
Anthurium hookeri	Bird's nest anthurium
Anthurium scherzerianum	Flamingo flower
Caladium spp.	Caladium
Calceolaria crenatiflora	Slipper wort
Catharanthus roseus	Madagascar periwinkle
Chirita lavandulacea	Hindustan gentian
Colummea hybrida	Gold fish plant
Davallia feieensis	Rabbit's foot fern
Epidendrum atropurpureum	Spice orchid
Episcia cupreata	Flame violet
Episcia dianthiflora	Lace – flower vine
Episcia reptans	Scarlet violet
Fittonia verschaffeltii	Red nerved fittonia
Fittonia verschaffeltii argyroneura	Silver nerved fittonia
Guzmania monostachia	Striped torch
Manettia inflata	Firecracker plant
Nephrolepis exaltata 'Bostoniensis'	Boston fern
Nephrolepis exaltata 'Fluffy Ruffles'	Fluffy ruffles
Pilea cadierei	Aluminium plant
Pilea microphylla	Artillery plant
Rhododendron hybrida	Azalea
Sinningia speciosa	Gloxinia
Soleirolia soleirolii	Baby tears

Plants which are classified under the category 2 (Average : 25 % to 49%)

Botanical Name	Common Name
Abutilon hybridum	Flowering maple
Acalypha hispida	Che nile plant
Achimene hybrida	Magic flower
Acorus calamus	Sweet flag
Acorus gramineus	Miniature sweet flag
Adiantum raddianum	Maiden hair fern
Adromischus cristatus	Crinkle-leaf plant
Adromischus festivus	Plover eggs
Aechme miniata 'Discolor'	Purplish coral berry
Aechmea 'Royal wine'	Royal wine bromeliad
Aeschynanthus marmoratus	Zebra basket vine
Aeschynanthus pulcher	Lipstick vine
Aglaonema modestum	Chinese evergreen
Aglaonema modestum 'Silver King'	Silver king
Alloplectus nummularia	Miniature pouch flower
Anthurium clarinervium	Dwarf crystal anthurium
Anthurium hookeri	Bird's nest anthurium
Anthurium scherzerianum	Flamingo flower
Aphelandra squarrosa	Zebra plant
Araucaria heterophylla	Norfolk island pine
Ardissa crenata	Ardisia
Asparagus densiflorus ' Myers'	Plume asparagus
Asparagus densiflorus ' Sprengeri'	Fax tail fern
Asparagus falcatus	Sickle thorn
Asplenium daucifolium	Mother fern
Asokebuyn budys	Bird's nest fern

Contd...

Botanical Name	Common Name
Begonia cubensis	Cuban holly
Begonia rex-cultorum	Rex begonia
Begonia semperflorens	Wax begonia
Billbergia nutans	Queens tears
Billbergia pyramidalis	Urn plant
Billbergia zebrina	Zebra plant
Brassaia actinophylla	Schefflera
Brassaia arboricola	Dwarf schefflera
Calathea insignis	Rattlesnake plant
Calathea makoyana	Peacock plant
Calathea roseopicta	Rose calathea
Callisia elegans	Striped inch plant
Carissa grandiflora 'Bonsai'	Bonsai natal plum
Carissa grandiflora 'Boxwood Beauty'	Boxwood beauty
Caryota mitis	Fish tail palm
Catharanthus roseus	Madagascar periwinkle
Ceropegia woodii	Rosary vine
Chamaerops humilis	European fan palm
Chirita lavandulacea	Hindustan gentian
Chlorophytum comosum 'Variegatum'	Variegated spider plant
Chlorophytum comosum 'vittatum'	Spider plant
Chrysalidocarpus lutescens	Areca palm
Cissus antartica	Kangaroo vine
Cissus rhombifolia	Grape leaf ivy
Cissus striata	Miniature grape ivy
Coleus blumei	Coleus
Colummea hybrida	Gold fish plant

Contd...

Botanical Name	Common Name
Crassula schmidtii	Red flowering crassula
Crossandra infundibuliformis	Crossandra
Cryptanthus fosteranus	Stiff pheasant leaf
Cryptanthus bivittatus 'Minor'	Dwarf rose stripe star
Cryptanthus zonatus	Zebra plant
Cyrtominum falcatum 'Rochfordianum'	House holly fern
Davallia feieensis	Rabbit's foot fern
Dieffenbachia 'Exotica Perfection'	Exotica perfection
Dizygotheca elegantissima	False aralia
Dracaena deremensis 'Janet Craig'	Janet craig
Dracaena deremensis 'Warneckii'	Warneckii
Dracaena fragrans 'Massangeana'	Corn plant
Dracaena marginata	Marginata
Dracaena surculosa	Gold dust dracaena
Epidendrum atropurpureum	Spice orchid
Epiphyllum hybrida	Golden pothos
Epipremnum aureum	Golden pothos
Epipremnum aureum 'Marble Queen'	Marble queen
Euphorbia coeralescens	Blue euphorbia
Euphorbia pulcherrima	Poinsettia
Euphorbia tirucalli	Milk bush
Fatshedera lizei	Botanical wonder plant
Ficus benjamina	Weeping fig
Ficus deltoidea	Mistletoe fig
Ficus elastic 'Decora'	Rubber plant
Ficus lyrata	Fiddle leaf fig
Ficus pumila 'Minima'	Dwarf creeping fig

Contd...

Botanical Name	Common Name
Ficus retusa	Cuban laurel
Ficus sagittata	Rooting fig
Ficus willdemaniance	Dwarf fiddle leaf fig
Gasteria hybrida	Ox tongue
Graptopetalum amethystinum	Jewel leaf plant
Gynura aurantiaca 'Purple Passion'	Purple passion
Hedera helix	English ivy
Hemigraphis alternata	Waffle plant
Hibiscus rosa-sinensis	Chinese hibiscus
Howea belmoreana	Belmore sentry palm
Howea forsterana	Kentia palm
Hoya carnosa 'Variegata'	Wax plant
Hoya kerrii	Sweetheart hoya
Hyacinthus orientalis	Hyacinth
Impatiens wallerana 'variegata'	Busy lizzie impatiens
Justicia brandegeana	Shrimp plant
Kalanchoe blossfeldiana	Christmas kalanchoe
Kalanchoe pumila	Dwarf purple kalanchoe
Kalanchoe tomentosa	Panda plant
Manettia inflata	Firecracker plant
Maranta leuconeura erythroneura	Red nerve plant
Maranta leuconeura kerchovina	Prayer plant
Mikania ternata	Plush vine
Nautilocalyx lynchii	Black alloplectus
Neoregelia carolinae 'Tricolor'	Tricolor bromeliad
Neoregelia spectabilis	Finger nail plant
Neoregelia zonata	Zonata

Contd...

Botanical Name	Common Name
Nephrolepis exaltata ' Bostoniensis'	Boston fern
Nidularium innocentii nana	Miniature bird's nest
Pachystachys clutea	Yellow shrimp plant
Paphiopedilum hybrida	Lady slipper orchid
Pedilanthus tithymaloides 'variegatus'	Devil's backbone
Pelargonium hortorum	House geranium
Pelargonium peltatum	Ivy geranium
Pellaea rotundifolia	Button fern
Oxalis flava	Finger oxalis
Oxalis rubra	Red oxalis
Pachyphytum oviferum	Pearly moonstones
Pachystachys clutea	Yellow shrimp plant
Paphiopedilum hybrida	Lady slipper orchid
Pedilanthus tithymaloides 'variegatus'	Devil's backbone
Pelargonium hortorum	House geranium
Pelargonium peltatum	Ivy geranium
Pellaea rotundifolia	Button fern
Pellionia pulchra	Satin pellionia
Peperomia crassifolia	Leather peperomia
Philodendron 'Emerald Queen'	Emerald queen
Philodendron 'Florida'	Florida
Philodendron scandens oxycardium	Heart – leaf philodendran
Philodendron selloum	Selloum
Phoenix roebelenii	Pigmy date palm
Pilea cadierei	Aluminium plant
Pilea microphylla	Artillery plant

Contd...

Botanical Name	Common Name
Plectranthus australis	Swedish ivy
Podocarpus macrophyllus	Podocarpus
Polyscias balfouriana 'Marginata'	Variegated balfour aralia
Polyscias fruticosa	Lady palm
Ruellia graeciznas	Red – spray ruellia
Saintpaulia hybrida	African violet
Saxifraga stolonifera	Strawberry geranium
Schlumbergera bridgesii	Christmas cactus
Sedum spectabile	Showy sedum
Sempervivum arachnoideum	Cow web house leek
Setcreasea pallida 'Purple Heart'	Purple heart
Sinningia speciosa	Gloxinia
Soleirolia soleirolii	Baby tears
Spathiphyllum 'Clevelandii'	Peace lily
Spathiphyllum 'Mauna Loa'	Mauna loa
Stapelia nobilis	Carrion flower
Strobilanthes dyeranum	Persian shield
Syngonium podophyllum	Nephthytis
Trillandsia bulbosa	Dancing bulb
Trillandsia lindenii	Blue – flowered torch
Tradescantia blossfeldiana	Flowering inch plant
Tradescantia sillamontana	White velvet plant
Vriesea splendens	Flaming sword
Zebrina pendula	Wondering jew

Plants which comes under the category 3 (Low : 5 % to 24%)

Botanical Name	Common Name
Astrophytum myriostigma	Bishop's cap
Cissus rotundifolia	Wax cissus
Crassula teres	Rattle snake tail
Echeveria agavoides	Molded wax
Echeveria elegans	Mexican snowball
Echinocereus reichenbachii	Lace cactus
Euphorbia coeralescens	Blue euphorbia
Euphorbia mileii	Crown-of-thorns
Haworthia cuspidata	Zebra haworthia
Haworthia subfasciata	Little zebra plant
Haworthia truncata	Clipped window plant
Hoya carnosa 'Variegata'	Wax plant
Kalanchoe pumila	Dwarf purple kalanchoe
Kalanchoe tomentosa	Panda plant
Opuntia vilis	Little tree cactus
Opuntia vulgaris	Irish mittens
Pachyphytum oviferum	Pearly moonstones
Pelargonium hortorum	House geranium
Pelargonium peltatum	Ivy geranium
Saintpaulia parva	Parva sansevieria
Saintpaulia trifasciata 'Hahnii'	Bird's nest sanseivieria
Saintpaulia trifasciata 'Laurentii'	Gold banded sansevieria
Sedum spectabile	Showy sedum
Sempervivum arachniodeum	Cow web house leek
Stapelia nobilis	Carrion flower
Yucca elephantipes	Spineless yucca

Watering

During summer months, plants require watering every day. Some plants like cacti and succulents should be watered if they show the sign of shrivelling. During winter months, due to cool temperature and dry air the media get dry up very quickly and many house plants probably die. So, watering should be done once in five to seven days.

Not all plants are similar in their water requirements. For example, a croton, which prefers high light, will likely need more frequent watering compared with a succulent plant such as Opuntia cactus. Both have similar light needs but dissimilar water requirements. The following table gives the idea of plants which require more or less water.

Plants Requires More Water	Plants Require Less Water
• Flowering plants	• Resting (or) dormant plants
• Plants potted in clay pots	• Recently repotted plant
• Plants grown in small pots	• Plants grown in high humidity
• Actively growing plants	• A plant located in a cool room
• Plants located in direct sun light	• Plants potted in non-porous containers
• Large leaved (or) thin leaved plants	• Plants with thick or rubbery leaves
• Plants that are native to wet areas	• Plants grown in a water retentive

One simple method to find the exact time for watering is, feel the soil by pushing a finger an inch or so below the surface. If the soil is still moist, no further water is needed. Water devices or water meters are also available to simplify watering. Based on which, watering may be given.

Improper watering causes many problems. Containers with saucers may cause an excessive build-up of soluble salts (from the applied fertilizer). High levels of soluble salts can cause damage to plant roots and a decline in growth. Discard any water that had drained in the saucer after irrigation, and apply large quantities of water to the soil to leach the accumulated soluble salts.

Over watering is the enemy for house plants, symptoms like fungus or mold in the soil surface, mushy brown roots at the bottom of the pot, standing water in the bottom of the container, young and old leaves falling off at the same time and leaves with brown rotten patches. Watering for house plant should be done based on the following ideas.

1. Keep media moist

2. Surface of media should dry before re-watering

3. Media can become moderately dry before re-watering

1. House plants which fit for the first category (Keep media moist)

Botanical Name	Common Name
Achimene hybrida	Magic flower
Acorus calamus	Sweet flag
Acorous gramineus	Miniature sweet flag
Adiantum raddianum	Maiden hair fern
Alloplectus nummularia	Miniature pouch flower
Anthurium clarinervium	Dwarf crystal anthurium
Anthurium hookeri	Bird's nest anthurium
Anthurium scherzerianum	Flamingo flower
Aphelandra squarrosa	Zebra plant
Araucaria heterophylla	Norfolk island pine
Ardissa crenata	Ardisia
Asplenium daucifolium	Mother fern
Asplenium nidus	Bird's nest fern
Caladium spp.	Caladium
Calathea insignis	Rattlesnake plant
Calathea makoyana	Peacock plant
Calathea roseopicta	Rose calathea
Calceolaria crenatiflora	Slipper wort
Chlorophytum comosum 'Variegatum'	Variegated spider plant
Chlorophytum comosum 'Vittatum'	Spider plant
Chrysalidocarpus lutescens	Areca palm
Colummea hybrida	Gold fish plant
Crossandra infundibuliformis	Crossandra
Davallia feieensis	Rabbit's foot fern
Epidendrum atropurpureum	Spice orchid

Contd...

Botanical Name	Common Name
Episcia cupreata	Flame violet
Episcia dianthiflora	Lace – flower vine
Episcia reptans	Scarlet violet
Fittonia verschaffeltii	Red nerved fittonia
Fittonia verschaffeltii argyroneura	Silver nerved fittonia
Fuchsia hybrida	Fuchsias
Hyacinthus orientalis	Hyacinth
Paphiopedilum hybrida	Lady slipper orchid
Pellaea rotundifolia	Button fern
Pilea cadierei	Aluminium plant
Pilea microphylla	Artillery plant
Saintpaulia hybrida	African violet
Soleirolia soleirolii	Baby tears
Spathiphyllum 'Clevelandii'	Peace lily
Spathiphyllum 'Mauna Loa'	Mauna loa

2. House plants which fit for the second category (Surface of media should dry before re-watering)

Botanical Name	Common Name
Abutilon hybridum	Flowering maple
Acalypha hispida	Che nile plant
Aechmea fasciata	Silver vase
Aechmea miniata 'Discolor'	Purplish coral berry
Aglaonema modestum	Chinese evergreen
Aglaonema modestum 'Silver King'	Silver king
Asparagus densiflorus 'Myers'	Plume asparaguas
Asparagus densiflorus 'Sprengeri'	Fox tail fern
Asparagus falcatus	Sickle thorn
Begonia cubensis	Cuban holly

Contd...

Botanical Name	Common Name
Begonia metallica	Metallic leaf ban begonia
Begonia rex-cultorum	Rex begonia
Begonia semperflorens	Wax begonia
Billbergia nutans	Queens tears
Billbergia pyramidalis	Urn plant
Billbergia zebrina	Zebra plant
Brassaia actinophylla	Schefflera
Brassaia arboricola	Dwarf schefflera
Carissa grandiflora 'Bonsai'	Bonsai natal plum
Carissa grandiflora 'Boxwood Beauty'	Boxwood beauty
Caryota mitis	Fish tail palm
Catharanthus roseus	Madagascar periwinkle
Ceropegia woodii	Rosary vine
Chamaerops humilis	European fan palm
Cissus antartica	Kangaroo vine
Cissus rhombifolia	Grape leaf ivy
Cissus striata	Miniature grape ivy
Coleus blumei	Coleus
Cordyline terminalis	Ti plant
Crassula argentea	Jade plant
Crassula lycopodioides	Toy cypress
Crassula schmidtii	Red flowering crassula
Cryptanthus bivittatus 'Minor'	Dwarf rose stripe star
Cryptanthus fosteranus	Stiff pheasant leaf
Cryptanthus zonatus	Zebra plant
Dieffenbachia 'Exotica Perfection'	Exotica perfection
Dizygotheca elegantissima	False aralia
Dracaena deremensis 'Janet Craig'	Janet craig

Contd...

Botanical Name	Common Name
Dracaena deremensis 'Warneckii'	Warneckii
Dracaena fragrans 'Massangeana'	Corn plant
Dracaena marginata	Marginata
Dracaena surculosa	Gold dust dracaena
Epipremnum aureum	Golden pothos
Epipremnum aureum 'Marble Queen'	Marble queen
Euphorbia coeralescens	Blue euphorbia
Euphorbia pulcherrima	Poinsettia
Euphorbia tirucalli	Milk bush
Fatshedera lizei	Botanical wonder plant
Ficus benjamina	Weeping fig
Ficus deltoidea	Mistletoe fig
Ficus elastica 'Decora'	Rubber plant
Ficus lyrata	Fiddle leaf fig
Ficus pumila 'Minima'	Dwarf creeping fig
Ficus retusa	Cuban laurel
Ficus sagittata	Rooting fig
Ficus willdemaniance	Dwarf fiddle leaf fig
Guzmania ligulata 'Major'	Scarlet star
Guzmania monostachia	Striped torch
Gynura aurantiaca 'Purple Passion'	Purple passion
Haworthia cuspidata	Zebra haworthia
Haworthia subfasciata	Little zebra plant
Hedera helix	English ivy
Hemigraphis alternata	Waffle plant
Hibiscus rosa-sinensis	Chinese hibiscus
Howea belmoreana	Belmore sentry palm
Howea forsterana	Kentia palm

Contd...

Botanical Name	Common Name
Hoya carnosa 'Variegata'	Wax plant
Hoya kerrii	Sweetheart hoya
Impatiens wallerana 'Variegata'	Busy lizzie impatiens
Justicia brandegeana	Shrimp plant
Kalanchoe blossfeldiana	Christmas kalanchoe
Maranta leuconeura erythroneura	Red nerve plant
Maranta leuconeura kerchovina	Prayer plant
Monstera deliciosa	Philodendron pertusum
Neoregelia carolinae 'Tricolor'	Tricolor bromeliad
Neoregelia spectabilis	Finger nail plant
Oxalis rubra	Red oxalis
Pachyphytum oviferum	Pearly moonstones
Pachystachys clutea	Yellow shrimp plant
Paphiopedilum hybrida	Lady slipper orchid
Pedilanthus tithymaloides 'Variegatus'	Devil's backbone
Pelargonium hortorum	House geranium
Pelargonium peltatum	Ivy geranium
Pellaea rotundifolia	Button fern
Pellionia pulchra	Satin pellionia
Peperomia caperata	Emerald ripple
Peperomia crassifolia	Leather peperomia
Peperomia obtusifolia	Baby rubber tree
Philodendron bipennifolium	Fiddle – leaf philodendron
Philodendron 'Emerald Queen'	Emerald queen
Philodendron 'Florida'	Florida
Philodendron scandens oxycardium	Heart leaf philodendron
Philodendron selloum	Selloum
Phoenix roebelenii	Pigmy date palm

Contd...

Botanical Name	Common Name
Platycerium bifurcatum	Staghorn fern
Plectranthus australis	Swedish ivy
Podocarpus macrophyllus	Podocarpus
Polyscias balfouriana 'Marginata'	Variegated balfour aralia
Polyscias fruticosa	Lady palm
Ruellia graeciznas	Red – spray ruellia
Sainpaulia parva	Parva sansevieria
Sainpaulia trifasciata 'Hahnii'	Bird's nest sanseivieria
Sainpaulia trifasciata 'Laurentii'	Gold banded sansevieria
Saxifraga stolonifera	Strawberry geranium
Sedum spectabile	Showy sedum
Sempervivum arachniodeum	Cow web house leek
Setcreasea pallid 'Purple Heart'	Purple heart
Sinningia speciosa	Gloxinia
Stapelia nobilis	Carrion flower
Streptocarpus hybridus	Cape primrose
Strobilanthes dyeranum	Persian shield
Syngonium podophyllum	Nephthytis
Trillandsia bulbosa	Dancing bulb
Trillandsia lindenii	Blue – flowered torch
Tolmiea menziesii	Piggyback plant
Tradescantia blossfeldiana	Flowering inch plant
Tradescantia sillamontana	White velvet plant
Vriesea splendens	Flaming sword
Yucca elephantipes	Spineless yucca
Zebrina pendula	Wondering jew

3. **House plants which fit for the third category (Media can become moderately dry before re-watering)**

Botanical Name	Common Name
Astrophytum myriostigma	Bishop's cap
Cissus rotundifolia	Wax cissus
Crassula falcata	Propeller plant
Crassula hemisphaerica	Arab'e turban
Crassula teres	Rattle snake tail
Echeveria agavoides	Molded wax
Echeveria elegans	Mexican snowball
Echinocereus reichenbachii	Lace cactus
Euphorbia coeralescens	Blue euphorbia
Gasteria hybrida	Ox tongue
Haworthia cuspidata	Zebra haworthia
Haworthia truncata	Clipped window plant
Kalanchoe pumila	Dwarf purple kalanchoe
Kalanchoe tomentosa	Panda plant
Mammillaria bocasana	Powder puff
Opuntia vilis	Little tree cactus
Opuntia vulgaris	Irish mittens
Saintpaulia parva	Parva sansevieria
Saintpaulia trifasciata 'Hahnii'	Bird's nest sanseivieria
Saintpaulia trifasciata 'Laurentii'	Gold banded sansevieria
Sedum spectabile	Showy sedum
Sempervivum arachniodeum	Cow web house leek
Stapelia nobilis	Carrion flower

Nutrition

There are three major constituents of plant food

- Nitrogen (N) - For leaf growth and 'greening' up of yellowing plants

- Phosphates (P_2O_5) - For root growth
- Potash (K_2O) - For flowers

Other elements of plant food will be the trace elements; these are generally present in most compound fertilizers. In general, plants grown in low light (interiors) will not require as much fertilizer as plants grown outside or in bright light. Always, liquid fertilizers holds good for the house plants. Less often is better than lots infrequently. Some plants such as ferns will need the feed diluted even further. Also, organic fertilizers (well rotten cowdung or oil cake) will give best results for the house plants. Moreover, the natural fertilizers are safer than the synthetics. A standard 20 : 20 : 20 formulation is fine for most indoor plants. Supplementing the inorganic nutrients with an organic amendment (liquid seaweed or fish emulsion, or a biostimulant), will provide some of the trace nutrients lacking in an inorganic fertilizer application. A top dressing of compost or worm castings is another effective way to add organic nutrients. For the plants which grow outdoors in homes, slow release nutrients will be found good over a long period of time. It can be mixed with the compost when potting up. The dose of fertilizers will be one teaspoon per pot.

Unknowingly, if synthetic fertilizers are used, the accumulation of soluble salts will be increased in course of time as a formation of crusty layer of salt deposits on the soil surface. In such a case, removal of salt deposits is highly essential, which can be achieved by leaching the soil with generous amounts of water. Because excessive salts can damage roots and makes the plant more susceptible to disease and insect attack.

Time for Nutrition Application

While placing the plant interiors, start fertilizing with low amount, then the dosage will be increased based on the sign of symptoms it shows. If overall plant colour becomes lighter, then increase the frequency of application. On the other hand, if the new growth is dark green, but the leaves are small and the space between the leaves seems longer than on the older growth, fertilize less often.

While watering the plants, nutrients in the soil get leached every time. Hence, it is most important to supplement the plants with nutrients once in a month especially when plants are in growing or flowering stage. But during severe winter, most of the plants will attain the dormant stage. Hence, it has to be withheld to maintain the healthiness of the plant.

When the plant starts to shed off their leaves, showing weak growth or overall appearance with yellow-green colour, will give an idea to fertilize the plants. It might also need more light or less water, so take the time to analyze all conditions before giving nutrition. If a plant is wilted, water well first then only nutrition should be given for getting fast recovery. Because, adding fertilizers when a plant does not need it can be worse than doing nothing for that.

Organic and Inorganic Nutrients

Organic nutrients

	Nitrogen (as N %)	Phosphorus (as P_2O_5 %)	Potassium (as K_2O %)
Cow manure	1.5	0.5	0.5
Farmyard manure	0.5	0.3	0.5
Compost	1.0	0.5	1.5
Cow-dung slurry from biogas plant	1.8	1.0	1.0
Sheep and goat droppings	3.0	1.0	2.0
Poultry litter	3.0	2.5	1.5
Vermicompost	1.5-2.0	1.0-1.5	0.60
Fishmeal	6.0	5.0	-
Bone meal	-	22.0	-
Blood meal	12.0 – 14.0	1.5	1.0
Oil cake (mustard)	5.0	1.5	1.0
Oil cake (Neem)	5.0	1.0	1.5
Leather meal	10.0	2.5	0.5
Stable manure (Manure from horse dung)	1.4	0.8	1.0
Chemical substances			
Urea	45	-	-
Ammonium sulphate	21	-	-
Calcium ammonium nitrate (CAN)	20	-	-
Di-ammonium phosphate (DAP)	18	46	-
Super phosphate	-	16	-
Sulphate of Potash (SOP)	-	-	50-54
Muriate of Potash (MOP)	-	-	60 (KCl)
Polyfeed (19 : 19 : 19)	19	19	19

Contd...

	Nitrogen (as N %)	Phosphorus (as P_2O_5 %)	Potassium (as K_2O %)
Sulphala (15 : 15 : 15)	15	15	15
Magnesium sulphate	-	-	Mg 20%
Boric acid	-	-	B 14%
Ferrous sulphate	-	-	Fe 20%
Ferric sulphate	-	-	Fe 20%
Zinc sulphate	-	-	Zn 36%

Acclimatization

Acclimatization is the adaptation of a plant to a new environment, and it is very important for the health and growth of indoor plants. Residential homes, with low-light interiors and low relative humidity, will most likely produce a stressful experience for plants the greater the difference between the previous environment and the environment of the house, the greater the stress the plant endures. To acclimatize plants at home, place newly purchased plants in bright areas for at least 3 to 4 weeks and then move them to their final location. Porches and patios are ideal bright places for your plants in the warm months, as long as the plants are not in direct sunlight. The most common symptom occurring in plants placed indoors is defoliation. As long as it is not extensive and it slows down after a few weeks, the plants will adjust to the particular location. Keep in mind, however, that each time the plant is moved around, it will experience an acclimatization period, and such changes may become evident.

Symptoms of acclimatized vs. nonacclimatized plants

Acclimatized Plants	Nonacclimatized Plants
Medium to dark green leaves	Yellowish to light green leaves
Large leaves	Small leaves
Flat leaves	Partially folder leaves
Thin leaves	Thick leaves
Widely spaced leaves	Closely spaced leaves
Long internodes	Short internodes
Thin to medium stems	Thick stems
Horizontal or slightly flexed leaf position	Upright leaf position
Few new leaves	Many new leaves
Wide branch angles	Acute angles

While purchasing, select only healthy looking plants with medium to dark green foliage (unless foliage is supposed to be a different color). Avoid plants with unnaturally spotted, yellow, or brown leaves. If the plant is unhealthy at the nursery, chances are that it will die soon after consumer purchase. Look for pests on the undersides of leaves. Remove the plant from the pot and examine the root system. Healthy roots generally are and should be visible along the outside of the soil ball and should have an earthy smell. Any discolorations, generally brown or blackened roots, are signs of problems. Some plants, such as Dracaenas, have roots with colours other than white. Unhealthy roots also may smell foul.

4

Propagation and Growing Media for House Plants

Propagation

- **Cuttings** – *Rex begonia, Gloxinia, Bryophyllum, Peperomia, African violet*
- **Root division** – Maranta, Dieffenbachia, Bellis, Gerbera, Kniphofia, Chrysanthemum
- **Ground layering** – Jasmine, Hydrangea, Azalea Climbing rose, *Vinca major*, Clematis, Hoya
- **Air layering** – Bougainvillea, Magnolia, Rhododendron, Croton, Coleus, Excoecaria, Ficus, Camellia, Hoya
- **Budding** - Rose
- **Stem cuttings** – Dieffenbachia, Acalypha, Aralia, Begonia, Dracaena, Coleus, Cordyline, Monstera, Polyscias, Crassula, Epipremnum, Excoecaria, Ficus, Fittonia, Gynura, Hedera, Irescine, Leea, Manihot, Oplismenus, Peperomia, Philodendron, Pleomele, Pseuderanthemum, Sanchezia, Schefflera, Scindapsus, Setcreasea,Tradescantia, Cebrina, Beloperone, Euphorbia, Gloxinia, Graptophyllum, Hoya, Hypoestes, Impatiens, Pelargonium, Aphelandra, Clerodendrum, Columnea, Episcia, Vanilla, Pinus, Juniperus, Thuja, Bambusa

- **Soft wood cuttings** – Coleus, Impatiens, Peperomia, Schefflera, Strobilanthes
- **Root cuttings** - Pleomele
- **Vine cutting** – Syngonium, Scindapsus, Philodendron
- **Leaf cuttings** – Peperomia, Scindapsus, Sansevieria, Croton, Gloxinia
- **Leaf blade cutting** - Bryophyllum
- **Runners** – Tradescantia, Zebrina, Philodendron, Epipremnum, Saxifraga
- **Division** – Asparagus, Aspidistra, Begonia, Calathea, Chlorophytum, Sansevieria, Cryptanthus, Cyperus, Fittonia, Helxine, Kaempferia, Maranta, Ophiopogon, Oplismenus, Pilea, Clematis, Gloxinia, Spathiphyllum, Hedychium, Heliconia, Oxalis, Bambusa, Orchids
- **Seed** – Asparagus, Begonia, Chlorophytum, Coleus, Ficus, Nolina, Pilea, Rhoeo, Gloxinia, Hypoestes, Impatiens, Senecio, Gloriosa, Hedychium, Palms, Cycas, Araucaria, Pinus, Juniperus, Thuja, Bambusa, Ravenala
- **Offsets** – Eucharis, Haemanthus, Hemarocallis, Strelitzia, Chlorophytum, Dieffenbachia, Nolina, Pandanus, Rhoeo, Biigergia, Neoregelia
- **Bulbils** – Furcraea, Cycas
- **Tissue culture** - Anthurium
- **Spores** - Aspidium
- **Suckers/Slips/by rooting leafy crown** - Ananas
- **Division of tubers** - Caladium
- **Tubers** - Gloriosa

Growing Media for House Plants

Media for house plants is somehow different from the plants grown in the ground. Because quantity of soil in the container is less which allows restricted root growth. While using soil, it must be heat sterilized by placing the soil in an oven at 90°C for at least 30 minutes. Coir (or) Cocopeat is used to increase aeration. Vermiculite and perlite are used to improve drainage in soil mixture. Perlite is recommended over vermiculite because it does not break down as easily. Charcoal is also added with the potting compost to absorb the foul gas of the medium and makes it favourable to the plants. Hence, the media suitable for house plants will fall in two categories *viz.*, soil based media and soilless media.

I. Formulation of Soil Based Root Media

Traditionally, a soil-based medium is made up of equal parts by volume of loam soil, river sand, and organic manure (vermicompost/FYM/Leaf mould) and adjusted to the proper pH level. The components of this type of root media are.

a. Field Soil

Field soil has a reasonable nutrient and water holding capacities. These two properties are significantly reduced if one third of the soil is replaced by sand since sand has very low nutrient and water holding capacities. To bring these properties at the desired level, organic manure is added into the medium at an expense of additional one third of the field soil. The other possible combination could be 3 : 2 : 1 : 1 of soil, organic manure, sand and cocopeat. This provides good water holding capacity along with high CEC value and a fair amount of aeration when it is coarse.

b. Sand

Sand is used in soil-based media to develop large diameter pores for good aeration. Two materials, perlite and polystyrene are the good substitutes for sand. A moist mixture of equal parts of soil, sand and peat moss (1 : 1: 1) weighs about 1,600 kg/m^3, which is suitable for use in greenhouses benches. It is not suitable for pot plants that are handled frequently or moved at greater distances, because of its weight.

II. Peat Moss and Peats

Peat is used as a soil modification. It is an important constituent of commercial components and act as a medium for plant growth.

a) Peat Moss

Peat moss is light tan or brown in colour and is least likely to decompose. The advantage of peat over manure or compost is that it is relatively stable (i.e., causes more permanent soil improvement), sterile, free from weeds, highly uniform and easy to mix with the soil. The recommended doses are 10-20 per cent by volume. They are formed by the accumulation of specific plant materials in poorly drained locations, It has a nitrogen content of 0.6-1.4 percent and decomposes slowly. It has high water holding capacity of up to 60 per cent of its volume in water.

b) Sphagnum Peat

It is the most popular form of organic matter for preparing root media. It consists of at least 75 per cent of partially decomposed stems and leaves of sphagnum moss. The peat neither adds appreciable amount of nutrients to the medium nor decreases the available amount of nutrients to the medium. It has pH level of 3.0-4.0 and requires 8-20 kg/m^2 of finely ground limestone to bring the pH up to a level that is best for most crops (5.5 – 6.5). The fibrous and cellular structure of the moss is recognizable and is free from decomposed colloidal residues, wood particles, silt and clay. They have good gas exchange capacity due to the presence of large pores between the aggregates.

d) Peat Humus

It is dark brown or black in colour, its water holding capacity is less than that of other peats, but is the most highly decomposable as compared to other peats. Its pH level ranges from 5.0-7.5. Nitrogen content of peat humus is moderately high, which makes it undesirable in seed-flat media or media used for salt sensitive plants. It is not commonly used in the greenhouses.

III. Soilless Media

a) Vermiculite

Vermiculite is an ore that has a dry bulk density of 880-1040 kg/m³, which reduces to 112-160 kg/m³ when expanded to a particular state used in root media. This lightweight property state is used in root media. This lightweight property makes it very useful in plant media. The water holding capacity of expanded vermiculite is high because of the large surface area within each particle. It has good aeration and drainage properties because of the presence of large pores between particles. There are numerous negative electrical charges on the surface of each vermiculite platelet which gives rise to a CEC of 1.9-2.7 me/100 cc. It is a secondary soil mineral that naturally tends to release magnesium and absorbs large amounts of potassium and ammonium. After high temperature treatment it becomes chemically inert and is not very stable physically. It is a desirable component of soil less media because of its high nutrient and water retention, good aeration and low bulk density. Under the weight of soil-based media, expanded Vermiculite tends to compress significantly thus reducing aeration. Vermiculite is generally not used with soils.

b) Sand

Sand is used in root media to induce proper drainage and aeration. Concrete grade sand (river sand) is commonly used in the media. Before using, it should be washed properly to be free from clay, silt and organic matter.

c) Perlite

Perlite is a silicaceous volcanic rock. it expands, forming white particles with numerous closed, air filled cells when crushed and heated to 982 °C. It is chemically inert, has negligible CEC (0.55me/100 cc) and is nearly neutral with a pH value of 7.5. Sand can be substitute by perlite to provide better aeration in root media. Its main advantage over sand is its lighter weight of about 96 kg/m³, as compared to 1,600-1,920 kg/m³ for sand.

Coconut Coir

It is a natural fibre extracted from the husk of coconut which has excellent water retention properties and also helps to circulate the air thoroughly within the pot or container. It comes in two forms *viz.*, roughly shredded or fine. The shredded form is good for combining with peat and compost, *etc*. Whereas, the fine form can be used by its own type or may be mixed with any media.

Gravel and Grit

Though, it will hold no water, it keeps the potting mixture as open and helps for free drainage. It can also add weight to containers for top heavy plants in order to stop them toppling over. Application of this gravel and grit as a layer will also aid in water retention and also arrest the rest of media from drying out. The composition of this while using for house plant as a media is less than 5%, but for plants like cacti, which dislike overwatering, its composition will be in higher level for quick drainage.

Media Composition of House Plants

Flowering plants	:	1 part garden loam (or) potting soil: 1 part sand (or) perlite (or) vermiculite: 1 part peat moss
Foliage plants	:	1 part garden loam (or) potting soil : 1 part sand (or) 2 parts peat moss
Cacti and succulents	:	2 parts garden loam (or) potting soil : 2 parts sand : 2 parts peat : 1 part perlite
Ferns	:	1 part garden loam (or) potting soil : 1 part peat moss : 1 part pine bark : 1 part coarse sand
Bromeliads	:	2 parts peat moss: 1part perlite (or) 1 part peat: 1 part fir bark (or) 1 part pine park
Orchids	:	3 parts osmunda tree fern fiber : 1 part redwood bark (or) 5 parts fir bark : 1 part perlite

5

Containers, Care and Management of House Plants

Selection of Containers

There are various types of pots available in the market for growing plants. Planters can enhance the decoration value of the plants. The following points are considered while selecting the container.

- Suitability for the plants needs
- Suitability for the needs of the individual and environment
- Cost and availability
- Strength and durability
- Drainage
- Weight

The style, shape, size of the container should complement the plants grown. Small containers are best for small, slow-growing plants, while fast-growing plants are better suited for large containers. Containers can be made from a wide range of materials – terra cotta, burnt clay, plastic (or) ceramic. In which, pots of burnt clay is widely used for exterior decoration. For indoor conditions, plastic pots are used widely. Now - a -days likely may be replaced by the introduction of grow bags which is also available in different sizes just like other types of pots. Clay pots can be glazed (or) unglazed.

The glazed pots restrict air exchange but offer more design choices. Unglazed pots evaporate water faster and plants which grow in this pot need more frequent watering. Disadvantages of clay containers are heavy weight and chance they will chip (or) break. But the materials like polyethylene, polyurethane, recycled plastic, fibreglass and plastic pots have the advantage of being light weight as well as chip and break resistant. Air exchange and water evaporation rates are generally lower in plastic containers compared with clay containers. Plants in plastic pots will not dry out as quickly as plants in clay pots, increasing the danger of over-watering.

The shape of pot may be round, oval, elliptical, cone-shaped, square, rectangular and oblong. Drainage hole should be placed in the pot is without drainage hole holds good for the plant like spathiphyllum which requires less quantity of water. Apart from this, glasses (or) tumbler shaped wooden vats are also available in the market. Moreover, a pot for orchids is totally different which contains more holes on the sides of the pot. The roots of orchids can absorb moisture and oxygen from atmosphere with a help of velamen tissue, as they are epiphytic.

Potting

Potting is the most important garden operation. Before potting, condition of the plant, size of the new pot should be taken into consideration. A pot should be appropriate to the size of the plant. Pots should be thoroughly cleaned. Then crocks are put over the drainage hole at the bottom. Then fill the bottom layer of the pot with sand, which will facilitate the passage of surplus water by percolation through the layer of aggregates. Plants should be placed in the centre of the pot and fill it with soil mixture just 1-2cm below (soil: sand: FYM @ 3 : 2 : 1) the rim of the pot gradually but slowly around the roots. By which the plant stands vertically. After filling the pot, it may be pressed with fingers to make it firm and compact. Then water the plants with rose can and place them in partial shade for proper establishment.

Repotting

When the root growth is noticed through the drainage hole (or) when the roots get matted around the outside of the earth-ball (or) when the soil ball of the pot becomes root bound, there is a need for repotting. While repotting, it is done to a pot of larger dimensions especially during rainy season which favours the formulation of new roots.

Plants like cacti, succulents and Aspidistra are slow growing and do not need repotting, while others like Begonia, Geranium, Syngonium and Chlorophytum are fast growing in nature, which needs repotting once in a year.

Pruning

a) Pruning of Plant Growth

When the plant is growing rapidly and you want to maintain a certain size, prune lightly and frequently, removing shoots or shoot tips when they are small. Examples are Jasmine, Rose, Bougainvillea and Geranium.

Remove dead flowers and leaves regularly. Leaves with tip and/or marginal necrosis, such as fluoride damage, should be trimmed to the healthy part. Examples are *Zebrina pendula* and *Scindapsus aureus*.

b) Root Pruning

Root pruning is a major operation, has to be done while repotting. In such case, one-third of the soil ball is reduced and fill the remain 2/3rd portion with fresh soil mixture. By using sharp scissors, long sharp knife, secateurs or a tensor saw, root pruning has to be done.

Pinching

It is the removal of apical shoots or tips of the plants. It helps to produce profuse branching like pruning. This is done regularly in coleus, geranium, chrysanthemum, tradescantia and other annual plants.

Cleaning

a) Cleaning the Containers

While re-using the containers, make sure that clean it well both inside and out by washing out any old compost, chemical, or paint residues. Clay pots often get a white crust on them after prolonged use. To remove this crust, scrub with a steel-wool pad or stiff brush in a vinegar and water solution. If the crust is thick, brush first with a dry steel wool pad. Rinse pots then soak them in a bleach solution (1 part bleach to 9 parts water) for 30 minutes. Rinse again. To remove salt and clinging earth from clay pots, rub with steel wool and diluted vinegar. Then soak pots in a bleach solution. Clean the plastic pots with a cloth dipped in warm soapy water. Scrub the pot until it's completely free of soil and grime. Soak the pot in a bleach solution as followed in clay pots.

b) Cleaning the Leaves

A clean plant is a healthy plant. Water flow causes salt accumulation along the leaf margins and/or tips, creating necrotic areas. Dust masks the normal leaf coloration, lessening the plant value. Dust on lower leaf surfaces may clog stomata (specialized cells involved in water transpiration), inhibiting gas exchange within the leaf. Leaves with thick, shiny cuticles (Croton, Ficus, Peace Lily, Bromeliads) should be cleaned with a damp sponge. Water should not be used when cleaning cacti, African violet leaves, and other plants with hairy leaves. Instead, use a clean, small paintbrush to remove dust.

Weeding

The weed growth is less while keeping the plants indoors. Anyhow, it is best to cover the soil surface with a layer of sphagnum moss which acts like a mulch and helps to suppress the weed growth.

Resting

Most of the house plants will undergo dormancy period. During that period, it shows the sign of die back (Gloxinia and tuberous rooted begonia) and shed off their leaves. Hence, watering is withheld to enjoy its resting stage. But evergreen plants do not have resting period. In case of winter flowering plants, they will enter the resting stage only after flowering is getting over.

6

House Plants - A Remedy for Air Pollutants

House (enclosed area) is a vulnerable place for noxious gases which can build up over time

Air Pollutants	Caused by
Benzene	A common solvent found in gasoline, oils, rubber, glues, paints, furniture wax, detergents and plastics
Formaldehyde	A water-soluble organic compound found in grocery bags, facial tissues, cleaning agents, emissions, disinfectants and fixatives or preservatives in consumer products and paper towels
Trichloroethylene	Industrial work- particularly in homes undergoing renovation (industrial product found in paints, varnishes, dry cleaning and adhesives)
Xylene	A chemical found in tobacco smoke, rubber, paint and vehicle exhaust
Xylene and toluene	A variety of household and consumer products
Ammonia	Aerosols and sprays used in the home as house cleaners, floor wax and fertilizers

House plants which help in removing air pollutants are

House Plants	Air Pollutants
Peace lily (*Spathiphyllum* spp.), Florist's chrysanthemum (*Chrysanthemum morifolium*)	Benzene, Formaldehyde, Trichloroethylene, Xylene and Toluene, Ammonia
Gerbera, Marginata (Dracaena), Peace lily, Janet craig, Bamboo palm	Trichloroethylene
Gerbera,English ivy, Peace lily, Warneckei (Dracaena), Bamboo palm	Benzene
Bamboo palm, *Dracaena deremansis*, Boston fern, Pot mum, Pygmy date palm	Formaldehyde
Pygmy date palm, Dumb cane, Marginata, Homalomena, Kimberley Queen Fern	Xylene
Broadleaf lady palm (*Rhapis excelsa*), Homalomena, Lily turf (*Liriope spicata*), Flamingo lily (*Anthurium andreanum*) and Pot mum	Ammonia
English ivy (*Hedera helix*), Variegated Snake plant (*Sansevieria trifasciata*)and Red edged dracaena (*Dracaena fragrans*)	Benzene, Formaldehyde, Trichloroethylene, Xylene and Toluene
Corn stalk dracaena (*Dracaena marginata*)	Benzene, Formaldehyde, Trichloroethylene
Broadleaf lady palm (*Rhapis excelsa*), Flamingo lily (*Anthurium andreanum*)	Trichloroethylene, Xylene and Toluene, Ammonia
Devil's ivy (*Epipremnum aureum*)	Benzene, Formaldehyde, Xylene and Toluene
Lily turf (*Liriope spicata*)	Formaldehyde, Trichloroethylene and Ammonia

Plants improve the air quality by combating volatile organic compounds (VOCs) commonly found in a home by absorbing VOCs through the pores on their leaves.

Effect of air pollutants (noxious gases) on human beings

Noxious Gas	Symptoms
Trichloroethylene	Excitement, dizziness, headache, nausea, vomiting followed by drowsiness and coma
Xylene	Irritation to mouth and throat, dizziness, headache, confusion, heart problem, liver and kidney damage and coma
Formaldehyde	Irritation to nose, mouth and throat, in severe cases , swelling of the larynx and lungs
Ammonia	Eye irritation, coughing and sore throat
Benzene	Irritation to eyes, drowsiness, dizziness, increase in heart rate, headaches, confusion, some cases can result in unconsciousness

Best air purifying house plants

Common Name	Botanical Name
Dwarf date palm	*Phoenix robelenii*
Boston fern	*Nephrolepis exaltata* Bostoniensis
Kimberley queen fern	*Nephrolepis obliterata*
Spider plant	*Chlorophytum comosum*
Bamboo palm	*Chamaedorea seifrizii*
Weeping fig	*Ficus benjamina*
Devils ivy	*Epipremnum aureum*
Flamingo lily	*Anthurium andreanum*
Lily turf	*Liriope spicata*
African Violet	*Saintpaulia* sp.
Heart leaf Philodendron	*Philodendron* sp.
Chinese Evergreen	*Aglaonema* sp., *Aglaonema Crispum* 'Deborah'
Aloe	*Aloe vera/A. barbadensis*
Peace lily	*Spathiphyllum* sp.
English ivy	*Hedera helix*
Janet Craig	*Draecana deremensis, Dracaena deremensis* 'Warneckii
Marginata or Dragon tree	*Dracaena marginata*
Corn Cane or Mass Cane	*Dracaena massangeana* or *Dracaena fragrans* Massangeana
Lady Palm	*Rhapis excelsa*
Indian rubber tree	*Sansevieria trifasciata*
Snake Plant	*Ficus elastica*
Pothos	*Epipremnum aureum* syn. *Scindapsus aureus*
Baby Rubber Plant	*Peperomia obtusifolia* or *Ficus robusta*
Gerbera Daisy	*Gerbera* sp. or *Gerbera jamesonii*
Schefflera, or Umbrella Tree	*Brassaia actinophylla* Syn: *Schefflera actinophylla*
Philodendron	*P. cordatum, P.scandens, P.oxycardium* or *P. selloum*

Contd...

Common Name	Botanical Name
Moth Orchid	*Phalaenopsis* spp.
Areca Palm	*Chrysalidocarpus lutescens*
Dwarf/Pygmy Date Palm	*Phoenix roebelenii*
Ficus alii	*Ficus maclellandii alii*
Mums	*Chrysanthemum* sp. or *Chrysanthemum morifolium*
Azalea	*Rhododendron simsii*

House plants which absorb carbon dioxide and release oxygen at night (the opposite of the process most plants follow).

1. Snake plant (*Sansevieria trifasciata* 'Laurentii')

2. Gerbera Daisy (*Gerbera* sp. or *Gerbera jamesonii*)

The above plants are highly suitable for bed room.

Plants which emit oxygen by removing toxin gas are

Heart Leaf Philodendron	*P.oxycardium*
Eucalyptus	*Eucalyptus* spp.
African Violets	*Saintpaulia ionantha*
Chinese Evergreen	*Aglaonema* sp., *Aglonema Crispum* 'Deborah'
Chrysanthemum	*Chrysanthemum morifolium*
Baby Rubber Plant	*Peperomia obtusifolia* or *Ficus robusta*
Weeping Fig or Ficus Tree	*Ficus benjamina*

7

Plant Protection Measures

I. Pests

House plants are less affected by pests and diseases as they are mostly hardy and slow growing. Common pests of house plants are red spider mites, aphids, mealy bugs, white flies, scale insects, thrips and caterpillars.

Red Spider Mites

They stick to the under surface of the leaves and suck the sap of the plant. They form delicate webs. House plants like *Azalea, Dracaena* are highly susceptible to this pest.

Aphids

Green or black aphids of small size remain adhere to the buds during cool weather. House plant like *Begonia* is highly susceptible to this pest.

Mealy Bugs

It forms a small white cottony patches on the under surface of the leaves of the affected plants chiefly along the leaf vines. House plants like *Begonia, Coleus, Poinsettia* and *Ficus elastica* are highly susceptible to this pest.

White Flies

These are very small flies in white colour which infest the house plants.

Scale Insects

Small round red or brown scales appear on the stems near the base. First they cover the basal portion and then spread upwards rapidly. The minute insects under the cover of the scales suck the sap of the plants. House plants like *Aspidistra, Cycas,* cacti, palm, fern and *Ficus elastica* are highly susceptible to this pest.

Thrips

These are very minute insects which are hardly visible. They destroy the growing shoots, which become brown and leaves distorted or crinkled.

Caterpillars

Soft-bodied long larvae eat into the leaves of house plants.

Control Measures

- While selecting the ornamental house plants from nursery, care should be taken to select a healthy plant.
- Avoid plants with unnaturally spotted, yellow, or brown leaves.
- If the plant is unhealthy at the nursery, chances are that it will die soon after purchase.
- Look for pests on the undersides of leaves.
- Remove the plant from the pot and examine the root system.
- Healthy roots generally should be visible along the outside of the soil ball and should have an earthy smell.
- Any discolorations, generally brown or blackened roots, are signs of problems.
- Some plants, such as Dracaenas, have roots with colors other than white. Unhealthy roots also may smell foul.
- If outdoor conditions permits, shift the affected houseplant outside in a protected area for few days, where natural predators will eventually come and kill the pest.
- Treat with insecticidal soap. Add 2 teaspoons of insecticidal soap for every 5 litres of water and wipe foliage and stems with the soapy water and soft cloth.
- Heavy infestations may be too difficult to treat. Discard these plants by burning and do not place them in the compost pile of home.

- Do not introduce beneficial insects indoors.
- Use Neemazal (Azadirachtin) @ 10,000 ppm at a dose of 1 ml/litre
- Use neemcake @ 50g/plant with the growing media against soil borne pests.
- Most pests can be controlled culturally on indoor plants without the use of chemicals.

II. Diseases

House plants are less prone to the attack by diseases. However, some of the common diseases seen in this type of house plants are leaf spot, stem rot, root rot, powdery mildew wilt, leaf fall and sunscald. Three factors are responsible for disease infection in plants.

They are: (1) a susceptible plant, (2) a viable pathogen, and (3) a favorable environment.

- Leaf spots are the most common problem, which showing symptoms of circular spots, black or brown patches especially in rainy season.
- Stem rot and root rot are highly infectious on cacti and succulents
- Powdery mildew cause the appearance of grayish white powdery coatings on the leaves and stems
- Wilting of leaves resulted due to lack of watering and also due to over watering
- Leaf fall may happen due to over watering or exposure to sudden chillness following hot hours
- Sun scald which causes pale brown patches on the leaf surface which occurs due to direct scorching sunlight

Control Measures

- Avoid creating stress to plants. A healthy plant is much more likely to fight off a disease than a stressed one.
- Soil-borne pathogens are commonly found on stressed plants. Soil-borne pathogens affect plants at or below the soil line; disease development is usually well underway before symptoms are noted on plant parts aboveground. Soil-borne diseases commonly occur when the growing medium is kept excessively moist and fertility levels are high. Low light and over-watering create favorable environments for soil-borne diseases under indoors conditions
- Fungal diseases are generally managed by the application of *Pseudomonas fluorescens* and *Trichoderma viridi*, biological control agents @ 10g/plant mixed with organic manures like vermicompost or FYM or Leafmould.

Annexure- Lists of House Plants

S. No.	Common Name	Botanical Name	Family
1.	Acalypha	*Acalypha godseffiana*	Euphorbiaceae
		Acalypha hispida	
		Acalypha sanderana	
		Acalypha sanderi	
		Acalypha wilkesiana 'Ceylon'	
		Acalypha wilkesiana 'Godseffiana'	
2.	-	*Achyranthes herbstii* syn. *Iresine herbstii*	Amaranthaceae
3.	Acorus	*Acorus gramineus*	Araceae
		Acorus gramineus 'Variegatus'	
4.	Aglaonema	*Aglaonema brevispathum* f. hospitum	
		Aglaonema commutatum	
		Aglaonema commutatum var. Elegans	
		Aglaonema commutatum 'Fransher'	
		Aglaonema commutatum 'Grafi'	
		Aglaonema commutatum 'Malay Beauty'	

Contd...

S. No.	Common Name	Botanical Name	Family
		Aglaonema commutatum 'Pseudobracteatum'	
		Aglaonema commutatum 'Treubii'	
		Aglaonema commutatum 'White Rajah'	
		Aglaonema costatum	
		Aglaonema costatum 'Foxii'	
		Aglaonema costatum forma immaculatum	
		Aglaonema crispum	
		Aglaonema modestum	
		Aglaonema nitidum	
		Aglaonema nitidum forma curtisii	
		Aglaonema oblongifolium	
		Aglaonema 'Parrot Jungle'	
		Aglaonema 'Pewter'	
		Aglaonema pictum	
		Aglaonema pseudobracteatum	
		Aglaonema roebelinii	
		Aglaonema siamense	
		Aglaonema 'Silver king'	
5.	-	*Aletris fragrans* syn. *Dracaena fragrans*	Asparagaceae
6.	Elephant ear plant	*Alocasia amazonica*	-
		Alocasia cucullata	
		Alocasia cuprea	
		Alocasia 'Hilo Beauty'	
		Alocasia indica	
		Alocasia jenningsii	
		Alocasia johnstonii	
		Alocasia lindenii	
		Alocasia lowii	

Contd...

S. No.	Common Name	Botanical Name	Family
		Alocasia macrorrhiza	
		Alocasia macrorrhiza 'Variegata'	
		Alocasia odora	
		Alocasia portei	
		Alocasia regina	
		Alocasia zebrina	
7.	Alpinia		Zingiberaceae
	Variegated ginger	*Alpinia* 'Sanderae'	
	-	*Alpinia speciosa*	
		Alpinia zerumbet	
		Alpinia zerumbet 'Variegata'	
8.	Copper leaf, Joseph's coat	*Alternanthera dentata*	Amaranthaceae
	-	*Alternanthera dentata* 'Rubiginosa'	
9.	Ananas	*Ananas comosus*	Bromeliaceae
	Variegated pineapple	*Ananas comosus* 'Variegatus'	
	-	*Ananas sativa*	
	-	*Ananas sativa* 'Variegatus'	
10.	-	*Anthericum* syn. *Chlorophytum*	Liliaceae
11.	Anthurium	*Anthurium andreanum*	Araceae
		Anthurium andreanum 'Rubrum'	
		Anthurium bakeri	
		Anthurium crassinervum syn. *Anthurium hookeri*	
		Anthurium crystallinum	
		Anthurium digitatum	
		Anthurium gracile	
		Anthurium hookeri	
		Anthurium macrolobium	

Contd...

S. No.	Common Name	Botanical Name	Family
		Anthurium magnificum	
		Anthurium olfersianum	
		Anthurium 'Robustum'	
		Anthurium venosum	
12.	Aralia		Araliaceae
	American spikenard, Angelica tree, Devil's walking stick, Herculis club, Udo & wild sarsaparilla	*Aralia balfouriana*	
		Aralia balfouriana sander 'Marginata'	
		Aralia elegantissima	
		Aralia filicifolia	
		Aralia fruticosa	
		Aralia guilfoylei	
		Aralia japonica	
		Aralia sieboldii	
		Aralia spectabilis	
13.	Araucaria		Araucariaceae
	Monkey puzzle tree, Norfolk island pink	*Araucaria columnaris*	
		Araucaria cookie syn. *Araucaria columnaris*	
		Araucaria cunninghamii	
		Araucaria excelsa	
		Araucaria heterophylla	
14.	-	*Areca lutescens* (syn. *Chrysalidocarpus lutescens*)	Arecaceae
15.	-	*Aregelia spectabilis* (syn. *Neoregelia spectabilis*)	-
16.	-	*Arum hederaceum* (syn. *Monstera deliciosa*)	Araceaea
17.	Asparagus		Liliaceae
	-	*Asparagus densiflorus*	
		Asparagus densiflorus 'Myers'	
		Asparagus densiflorus 'Sprengeri'	

Contd...

S. No.	Common Name	Botanical Name	Family
	Sickle thorn	*Asparagus falcatus*	
	-	*Asparagus myersii* (syn. *Asparagus densiflorus* 'Myers')	
		Asparagus plumosus (syn. *Asparagus setaceus*)	
		Asparagus plumosus 'Nanus'	
		Asparagus racemosus	
		Asparagus setaceus	
		Asparagus setaceus 'Pyramidalis'	
		Asparagus sprengeri (syn. *Asparagus densiflorus* 'Sprengeri')	
		Asparagus tetragonus (syn. *Asparagus racemosus*)	
18.	Aspidistra		Liliaceae
	Bar room plant, Parlour palm, Cast iron plant	*Aspidistra elatior*	
	Variegated cast iron plant	*Aspidistra elatior* 'Variegata'	
		Aspidistra lurida 'Variegata' (syn. *Aspistra elatior*)	
19.	Bambusa	*Bambusa glaucescens*	Poaceae (Graminae)
	-	*Bambusa multiplex* (syn. *Bambusa glaucescens*)	
	Budha's bamboo, Budha's belly bamboo	*Bambusa ventricosa*	
20.	Begonia		Begoniaceae
	Star begonia, star leaf begonia	*Begonia heracleifolia*	
	-	*Begonia jatrophaefolia* (syn. *Begonia heracleifolia*)	
		Begonia ramentacea	
		Begonia rex	
		Begonia rex 'Salamander'	
	King begonia, painted leaf begonia	*Begonia rex* 'Silver Queen'	
		Begonia speculata	

Contd...

S. No.	Common Name	Botanical Name	Family
21.	Beloperone		Acanthaceae
	False hop, shrimp plant	*Beloperone guttata*	
		Beloperone guttata 'Yellow queen'	
22.	Vase plant	*Billbergia*	Bromeliaceae
	Marbled rainbow plant	*Billbergia* 'Fantasia'	
	Fool proof plant, summer torch	*Billbergia pyramidalis*	
	-	*Billbergia pyramidalis* var. concolor	
		Billbergia saundersii	
		Billbergia thyrsiflora	
		Billbergia thyrsoidea (syn. *Billbergia pyramidalis* var. concolor)	
23.	Australian ivy palm, Australian umbrella tree, Octopus tree, Queens land umbrella tree, Starleaf	*Brassaia actinophylla*	Araliaceae
24.	Mother- in- law plant, Elephant's ear, Angel-wings	*Caladium argyrites* (syn. *Caladium humboldtii*)	Araceae
	-	*Caladium humboldtii*	
25.	Calathea	*Calathea bachemiana*	Marantaceae
	-	*Calathea bachemiana* 'Trifasciata' (syn. *Calathea bachemiana*)	
		Calathea bella	
		Calathea discolour	
		Calathea flavescens	
		Calathea grandiflora	
		Calathea kegeliana	
		Calathea lietzei	
		Calathea louisae	
		Calathea medio- picta	

Contd...

S. No.	Common Name	Botanical Name	Family
		Calathea ornata	
		Calathea ornata var. roseo- lineata	
		Calathea ornata var. 'Sanderana'	
		Calathea princeps	
		Calathea pulchella (syn. *Calathea zebrina* 'Humilior'*)*	
		Calathea sanderana (syn. *Calathea ornata* 'Sanderana'*)*	
		Calathea tigrina (syn. *Calathea zebrina* 'Humilior'*)*	
		Calathea undulata	
		Calathea zebrina	
		Calathea zebrina 'Humilior'	
26.	Callisia		Commelinaceae
	-	*Callisia elegans*	
		Callisia fragrans	
27.	-	*Carex variegata* (syn. *Acorus gramineus* 'Variegatus'*)*	Araceae
28.	Carludovica	*Carludovica palmata*	Cyclanthaceae
29.	Caryota		Arecaceae (Palmae)
	Burmese fish tail palm, clustered fish tail palm, tufted fish tail palm	*Caryota mitis*	
	Fish tail palm, Indian sago, Jaggery palm, Kitool palm, Kittul tree, Malabar sago palm, sago palm, toddy palm, Wine palm	*Caryota urens*	
30.	-	*Chamaerops biroo* (syn. *Livistona rotundifolia)*	Arecaceae
	-	*Chamaerops byrrho* (syn. *Livistona rotundifolia)*	
	-	*Chamaerops excelsa* (syn. *Rhapis excelsa)*	Araceae

Contd...

S. No.	Common Name	Botanical Name	Family
31.	Chlorophytum		Liliaceae
	Saint Bernard's lily	*Chlorophytum comosum*	
	-	*Chlorophytum comosum* 'Mandaianum'	
		Chlorophytum comosum 'Variegatum'	
		Chlorophytum comosum 'Vittatum'	
		Chlorophytum elatum	
		Chlorophytum elatum 'Vittatum'	
		Chlorophytum mandaianum (syn. *Chlorophytum comosum* 'Mandaianum'*)*	
		Chlorophytum sternbergianum (syn. *Chlorophytum comosum)*	
		Chlorophytum vittatum (syn. *Chlorophytum comosum* 'Vittatum'*)*	
32.	Chrysalidocarpus		Arecaceae
	Areca palm, Butterfly palm, Cane palm, Golden feather palm, Yellow butterfly palm, yellow palm	*Chrysalidocarpus lutescens*	
33.	Cissus		Vitaceae
	Princess vine	*Cissus sicyoides*	
34.	Clusia		Clusiaceae (Guttiferae)
	Balsam apple, Balsam tree, Copey, Fat pork tree, Pitch apple cupey	*Clusia rosea*	
35.	-	*Coccoloba platyclada* (syn. *Homalocladium platycladium)*	-
36.	Codiaeum		Euphorbiaceae
	Croton, variegated laurel	*Codiaeum variegatum*	
37.	Coleus		Lamiaceae
	Basket coleus	*Coleus pumilus*	
	-	*Coleus rehneltianus* (syn. *Coleus pumilus)*	

Contd...

S. No.	Common Name	Botanical Name	Family
38.	Colocasia		Araceae
	Elephant ear plant	*Colocasia affinis*	
	-	*Colocasia mafaffa* (syn. *Xanthosoma mafaffa*)	
39.	-	*Commelina zebrina* (syn. *Xanthosoma mafaffa*)	Commelinaceae
40.	Cordyline	*Cordyline angusta*	Agavaceae
		Cordyline australis 'Atropurpurea'	
		Cordyline australis	
		Cordyline australis 'Cuprea'	
		Cordyline australis 'Doucetii'	
		Cordyline baptisii (syn. *Cordyline terminalis* 'Baptisii')	
		Cordyline rumphii (syn. *Dracaena hookerana*)	
	Good –luck plant, Hawaiian good – luck plant, Ti, Tree - of kings	*Cordyline terminalis*	
	-	*Cordyline terminalis* 'Angusta'	
		Cordyline terminalis 'Baby Ti'	
		Cordyline terminalis 'Baptisii'	
		Cordyline terminalis 'Bicolor'	
		Cordyline terminalis 'Fire band'	
	Flaming dragon tree	*Cordyline terminalis* 'Madame Eugene andre'	
	-	*Cordyline terminalis* 'Margaret storey'	
		Cordyline terminalis var. minima (syn. *Cordyline terminalis* 'Baby Ti')	
	Hawaiian good –luck plant, Tree - of kings	*Cordyline terminalis* 'Ti'	
41.	Costus		Zingiberaceae
	Crape ginger or Malay ginger, Spiral ginger	-	
	Spiral ginger, wild ginger	*Costus speciosus*	

Contd...

S. No.	Common Name	Botanical Name	Family
42.	Crinum		Amaryllidaceae
	Crinum lily, milk and wine lily, poison bulb, spider lily	-	
	Poison bulb	*Crinum asiaticum* 'Variegatum'	
43.	Croton	*Croton bicolour (syn. Excoecaria bicolour)*	Euphorbiaceae
44.	-	*Cryptophragmium ceylanicum (syn. Gymnostachyum ceylanicum)*	Acanthaceae
45.	Ctenanthe	*Ctenanthe humilis*	Marantaceae
46.	Curculigo		Hypoxidaceae
	Palm grass	*Curculigo capitulata*	
47.	-	*Curmeria wallisii (syn. Homalomena wallisii)*	Araceae
48.	-	*Cyanotis vittata (syn. Zebrina pendula)*	Commelinaceae
49.	Sago palm, Australian nut palm	*Cycas*	Cycadaceae
	Japanese fern palm, Japanese sago palm, Sago palm	*Cycas revoluta*	
	Bread palm	*Cycas rumphii*	
50.	Galingale, Umbrella sedge	*Cyperus*	Cyperaceae
	Broad leaf umbrella plant	*Cyperus albostriatus*	
	Umbrella palm, Umbrella plant, Umbrella sedge	*Cyperus alternifolius*	
	-	*Cyperus alternifolius* 'Variegatus'	
	-	*Cyperus diffusus (syn. Cyperus albostriatus)*	
	-	*Cyperus elegans (syn. Cyperus albostriatus)*	
	-	*Cyperus laxus (syn. Cyperus albostriatus)*	
51.	Cyrtanthera		
	-	*Cyrtanthera magnifica (syn. Justicia carnea)*	Acanthaceae
		Cyrtanthera pohliana (syn. Justicia carnea)	

Contd...

S. No.	Common Name	Botanical Name	Family
52.	Cyrtosperma	*Cyrtosperma johnstonii*	Araceae
53.	Dichorisandra		Commelinaceae
	-	*Dichorisandra musaica*	
		Dichorisandra reginae	
54.	Dieffenbachia		Araceae
	Dumb cane, Mother – in- law plant	*Dieffenbachia amoena*	
	-	*Dieffenbachia amoena* 'Hi-color' *(syn. Dieffenbachia amoena* 'Tropic snow'*)*	
		Dieffenbachia amoena 'Tropic snow'	
		Dieffenbachia amoena 'Tropical Topaz' *(syn. Dieffenbachia amoena* 'Tropic snow'*)*	
		Dieffenbachia 'Arvida' *(syn. Dieffenbachia* 'Exotica'*)*	
		Dieffenbachia baraquiniana (syn. Dieffenbachia maculata 'Baraquiana'*)*	
		Dieffenbachia baumanii (syn. Dieffenbachia seguine var. *irrorata)*	
		Dieffenbachia bausei	
		Dieffenbachia brasiliense (syn. Dieffenbachia maculata)	
		Dieffenbachia 'Exotica'	
		Dieffenbachia fournieri	
		Dieffenbachia gigantea (Dieffenbachia maculata 'Baraquiniana'*)*	
		Dieffenbachia irrorata (Dieffenbachia seguine var. irrorata*)*	
		Dieffenbachia jenmannii (Dieffenbachia maculata 'Jenmannii'*)*	
		Dieffenbachia 'Leonii'	
		Dieffenbachia 'Leopoldii'	
	Spotted dumb cane	*Dieffenbachia maculata*	
	-	*Dieffenbachia maculata* 'Baraquiniana'	
		Dieffenbachia maculata 'Jenmannii'	

Contd...

S. No.	Common Name	Botanical Name	Family
		Dieffenbachia maculata 'Viridis'	
		Dieffenbachia memoria- corsii	
		Dieffenbachia oerstedii	
		Dieffenbachia picta	
		Dieffenbachia picta baraquiniana	
		Dieffenbachia picta var. jenmannii	
		Dieffenbachia picta var. memoria	
		Dieffenbachia picta 'Rudolph Roehrs'	
		Dieffenbachia picta 'Viridis' *(syn. Dieffenbachia maculata* 'Viridis'*)*	
		Dieffenbachia regina	
		Dieffenbachia roehrsii (syn. Dieffenbachia maculata 'Rudolph Roehrs'*)*	
		Dieffenbachia sanderae	
		Dieffenbachia seguine	
		Dieffenbachia shutteworhiana	
		Dieffenbachia splendens	
		Dieffenbachia verschaffeltii (syn. Dieffenbachia maculata 'Baraquiniana'*)*	
55.	Dizygotheca	*Dizygotheca elegantissima*	Araliaceae
56.	Dorstenia	*Dorstenia contrajerva*	Moraceae
		Dorstenia tureraefolia	
57.	Dracaena		Asparagaceae
	-	*Dracaena angustifolia (syn. Dracaena reflexa* 'Angustifolia'*)*	
		Dracaena baptisii (syn. Cordyline terminalis 'Baptisii'*)*	
		Dracaena calocoma (syn. Cordyline australis 'Aureostriata'*)*	
		Dracaena cineta	
	Striped Dracaena	*Dracaena deremensis* 'Warneckii'	
	Corn plant	*Dracaena fragrans*	
	-	*Dracaena fragrans* 'Lindenii'	
	Corn stalk plant	*Dracaena fragrans* 'Massangeana'	

Contd...

S. No.	Common Name	Botanical Name	Family
	Painted dragon lily	*Dracaena fragrans* 'Victoria'	
	-	*Dracaena godseffiana* (*Dracaena surculosa*)	
		Dracaena godseffiana 'Kelleri' (*Dracaena surculossa* 'Kelleri')	
	Queen of dracaenas	*Dracaena goldieana*	
	-	*Dracaena gracilis*	
	Leather dracaena	*Dracaena hookerana*	
	-	*Dracaena indivisa* (syn. *Cordyline australis* 'Aureostriata')	
		Dracaena marginata (syn. *Dracaena cincta*)	
		Dracaena reflexa	
		Dracaena reflexa 'Angustifolia'	
	Song of India	*Dracaena reflexa* 'Variegata'	
	-	*Dracaena rumphii* (syn. *Dracaena hookerana*)	
		Dracaena surculosa	
		Dracaena surculosa 'Kelleri'	
		Dracaena surculosa 'Punctulata'	
		Dracaena terminalis	
		Dracaena thalioides	
		Dracaena umbraculifera	
58.	-	*Dracontium pertusum* (syn. *Monstera deliciosa*)	Araceae
59.	-	*Drejerella guttata* (syn. *Beloperone guttata*)	Acanthaceae
60.	Drimiopsis	*Drimiopsis kirkii*	Liliaceae
61.	Golden dewdrop, Pigeon berry, sky flower	*Duranta*	Verbenaceae
	Brazilian sky flower, Creeping skin flower, Golden dewdrop, Pigeon berry, sky flower	*Duranta repens* 'Variegata'	
62.	Encephalartos	*Encephalartos villosus*	Zamiaceae

Contd...

S. No.	Common Name	Botanical Name	Family
63.	Ensete		Musaceae
	Abyssinia banana	*Ensete ventricosum*	
64.	-	*Ephemerum bicolour (syn. Rhoeo spathacea)*	Commelinaceae
65.	Phothos vine, Ivy arum	*Epipremnum*	Araceae
	Devil's Ivy,Golden Ceylon creeper, Golden hunter's robe, Golden pothos, Hunter's robe, Ivy arum, Money plant, Pothos, Pothos vine, Solomon island ivy, Taro vine, Variegated philodendron	*Epipremnum aureum*	
	Taro vine	*Epipremnum aureum* 'Marble queen'	
	Golden pothos	*Epipremnum aureum* 'Wilcoxii'	
	-	*Epipremnum pinnatum (syn. Epipremnum aureum)*	
66.	Eranthemum	*Eranthemum albomarginatum*	Acanthaceae
	-	*Eranthemum atrosanguineum (syn. Pseuderanthemum nigrum)*	
		Eranthemum bicolour (syn. Pseuderanthemum bicolor)	
		Eranthemum laxiflorum (syn. Pseuderanthemum laxiflorum)	
		Eranthemum nigrum (syn. Pseuderanthemum nigrum)	
		Eranthemum reculatum (syn. Pseuderanthemum reculatum)	
		Eranthemum rubronervum (syn. Fittonia verschaffeltii var. verschaffeltii)	
		Eranthemum rubrovenosum (syn. Fittonia verschaffeltii var. verschaffeltii)	
		Eranthemum schomburgkii (syn. Pseuderanthemum reticulatum)	
		Eranthemum tricolor (syn. Pseuderanthemum atropureum 'Tricolor'*)*	

Contd...

S. No.	Common Name	Botanical Name	Family
67.	Eucharis	*Eucharis amazonica* (syn. *Eucharis grandiflora*)	Amaryllidaceae
		Eucharis grandiflora	
68.	Spindle tree, straw berry bush, wahoo	*Euonymus*	Celastraceae
	Japanese spindle tree, spindle tree	*Euonymus japonicus*	
		Euonymus japonica var. *medio picta*	
	Gold heart Euonymus	*Euonymus japonicus* 'Media- pictus'	
69.	Eurycles	*Eurycles amboinensis*	Amaryllidaceae
	Brisbane lily	*Eurycles sylvestris*	
	-	*Eurycles amboinensis*	
70.	Excoecaria	*Excoecaria bicolor*	Euphorbiaceae
71.	Fatsia		Araliaceae
	False castor oil plant, Fig leaf palm, Formosa rice tree, Glossy leaved paper plant, Japanese aralia, Japanese fatsia, Japanese fig, Paper plant	*Fatsia japonica*	
72.	Ficus		Moraceae
	-	*Ficus belgica* (syn. *Ficus elastica*)	
		Ficus benjamina var. *nuda*	
		Ficus comosa (syn. *Ficus benjamina* var. *nuda*)	
		Ficus decora (syn. *Ficus elastica* 'Decora')	
	Assam rubber, India rubber tree, Indian rubber plant, Rubber plant	*Ficus duvivieri* (syn. *Ficus elastica*)	
	Broad leaved Indian rubber plant, wide leaf rubber plant	*Ficus elastica* 'Decora'	
	Black prince	*Ficus elastica* 'Rubra'	

Contd...

S. No.	Common Name	Botanical Name	Family
	Variegated rubber plant	*Ficus elastica* 'Variegata'	
	Chinese fig, fiddle leaf, fiddle leaved fig	*Ficus lyrata*	
		Ficus pandurata (syn. *Ficus lyrata*)	
	Climbing fig, creeping fig, Indian Ivy, Creeping rubber plant	*Ficus pumila*	
	-	*Ficus repens* (syn. *Ficus pumila*)	
		Ficus rubra (syn. *Ficus elastica*)	
		Ficus stipulata (syn. *Ficus pumila*)	
73.	Fittonia		Acanthaceae
	-	*Fittonia argyroneura* (syn. *Fittonia verschaffeltii* var. *argyroneura*)	
		Fittonia pearcei (syn. *Fittonia verschaffeltii* var. *pearcei*)	
		Fittonia rubronervum (syn. *Fittonia verschaffeltii* var. *verschaffeltii*)	
		Fittonia rubronervosa (syn. *Fittonia verschaffeltii* var. *verschaffeltii*)	
		Fittonia verschaffeltii var. *argyroneura*	
		Fittonia verschaffeltii var. *pearcei*	
	Mosaic plant, Nerve plant	*Fittonia verschaffeltii* var. *verschaffelti*	
74.	Geogenanthus		Commelinaceae
	Seersucker plant	*Geogenanthus undatus*	
75.	Globba	*Globba schomburgkii*	Zingiberaceae
76.	Gloxinia	*Gloxinia maculata* (syn. *Gloxinia perennis*)	Gesneriaceae
		Gloxinia perennis	
77.	Graptophyllum	*Graptophyllum gilligani*	Acanthaceae
		Graptophyllum pictum	
		Graptophyllum picturatum	
78.	Gymnostachyum	*Gymnostachyum ceylanicum*	Acanthaceae
		Gymnostachyum pearcei (syn. *Fittonia verschaffeltii* var. *pearcei*)	
		Gymnostachyum verschaffeltii (*Fittonia verschaffeltii* var. *verschaffeltii*)	

Contd...

S. No.	Common Name	Botanical Name	Family
79.	Gynura		Asteraceae
	Purple velvet plant, Royal velvet plant, Velvet plant	*Gynura aurantiaca*	
	-	*Gynura bicolor*	
		Gynura procumbens	
		Gynura sarmentosa (syn. *Gynura procumbens*)	
80.	Ginger lily	*Hedychium*	Zingiberaceae
	Butterfly ginger, butterfly lily, Cinnamon jasmine, Garland flower, Ginger lily, Indian garland flower	*Hedychium coronarium*	
	Yellow ginger	*Hedychium flavum*	
81.	Heliconia	*Heliconia aureostriata*	Heliconiaceae
	Baliscer, False plantain, Fire bird, Macaw flower, Wild plantain	*Heliconia bihai*	
	-	*Heliconia* 'Frotsy' (*Heliconia metallica*)	
		Heliconia indica var. *aureo- striata* (*Heliconia striata*)	
		Heliconia metallica	
		Heliconia rubro-striata (syn. *Heliconia illustris*)	
		Heliconia striata	
82.	Hemigraphis		Acanthaceae
	Red Ivy, Red flame ivy	*Hemigraphis alternata*	
	-	*Hemigraphis colorata* (syn. *Hemigraphis alternata*)	
	Purple waffle plant	*Hemigraphis* 'Exotica'	
83.	-	*Heptapleurum venulosum* (syn. *Schefflera venulosa*)	Araliaceae
84.		*Heteropsis pertusa* (syn. *Monstera deliciosa*)	Araceae
85.		*Hippobroma longiflora*	Lobeliaceae
86.		*Hoffmannia refulgens*	Rubiaceae

Contd...

S. No.	Common Name	Botanical Name	Family
87.	Homalocladium		Polygonaceae
	Centipede plant, Ribbon bush, Tapeworm plant	*Homalocladium platycladum*	
88.	Homalomena		Araceae
	-	*Homalomena lindenii*	
		Homalomena rubescens	
89.	Porcelain flower, Wax plant, Wax vine	*Hoya*	Apocynaceae
	Variegated wax plant	*Hoya carnosa*	
	Silver pink vine	*Hoya purpureo fusca*	
90.	Freckle face	*Hypoestes*	Acanthaceae
	-	*Hypoestes phyllostachya*	
		Hypoestes rotundifolia	
		Hypoestes sanguinolenta	
91.	Blood leaf	*Iresine*	Amaranthaceae
	Beef plant, Beef steak plant, Chicken gizzard	*Iresine herbstii*	
		Iresine reticulara (syn. *Iresine herbstii*)	
92.	-	*Isotoma longiflora* (syn. *Hippobroma longiflora*)	Campanulaceae
93.	Justicia	*Justicia brandegeana* (syn. *Beloperone guttata*)	Acanthaceae
	Brazilian plume, Flamingo plant, Kings crown, Paradise plant, Plume flower, Plume plant	*Justicia carnea*	
	-	*Justicia gilligani* (syn. *Graptophyllum gilligani*)	
		Justicia magnifica (syn. *Justicia carnea*)	
		Justicia picta (syn. *Graptophyllum pictum*)	
94.	Kaempferia		Zingiberaceae
	Variegated gingerlily	*Kaempferia gilbertii*	
	-	*Kaempferia involucrata*	
	Dwarf ginger lily, Peacock lily	*Kaempferia roscoeana*	
	-	*Kaempferia rotunda*	

Contd...

S. No.	Common Name	Botanical Name	Family
95.	-	*Latania borbonica* (syn. *Livistona chinensis*)	Arecaceae
96.	-	*Laurentia longiflora* (syn. *Hippobroma longiflora*)	Campanulaceae
97.	Leea		Vitaceae
	West Indian holly	*Leea coccinea*	
98.	Lily turf	*Liriope*	Liliaceae
	-	*Liriope graminifolia* var. *densiflora* (syn. *Liriope muscari*)	
		Liriope japonica (syn. *Ophiopogon japonicus*)	
	Big blue lily turf, Turf lily	*Liriope muscari*	
99.	Livistona		Arecaceae
	-	*Livistona altissima* (*Livistona rotundifolia*)	
	Chinese fan palm, Chinese fountain palm	*Livistona chinensis*	
	-	*Livistona mauritiana* (syn. *Livistona chinensis*)	
	Fan palm	*Livistona rotundifolia*	
100.	Cassava, Manioc, Tapioca plant (or) Yuco	*Manihot*	Euphorbiaceae
	Cassava (or) Tapioca plant, Variegated Tapioca plant	*Manihot esculenta* 'Variegata'	
	-	*Manihot utilissima* 'Variegata' (*Manihot esculenta* 'Variegata')	
101.	Arrow root, prayer plant	*Maranta*	Marantaceae
	Arrow root, Bermuda arrow root, Obedience plant	*Maranta arundinacea*	
	-	*Maranta arundinacea* 'Variegata'	
		Maranta bella	
		Maranta conspicua	
		Maranta kegeljanii	
		Maranta lietzei	
		Maranta undulata (syn. *Calathea undulata*)	
		Maranta zebrina (syn. *Calathea zebrina*)	

Contd...

S. No.	Common Name	Botanical Name	Family
102.	-	*Molineria recurvata (syn. Curuligo capitulata)*	Hypoxidaceae
103.	-	*Mondo jaburan* (syn. *Ophiopogon jaburan)*	Asparagaceae
104.	-	*Mando japonicum* (syn. *Ophiopogon japonicus)*	
105.	Monstera		Araceae
	Bread fruit vine, Ceriman, Cut leaf philodendron, Gruyere cheese plant, Fruit salad plant, Hurricane plant, Mexican bread fruit, Once seen never forgotten plant, Split leaf philodendron, Swiss cheese plant & Window plant	*Monstera deliciosa*	
	-	*Monstera deliciosa* 'Albo-variegata'	
	Variegated philodendron	*Monstera deliciosa* 'Marmorata'	
	-	*Monstera epipremnoides*	
		Monstera falcifolia	
		Monstera guttifera	
		Monstera leichtlinii	
		Monstera lennea	
		Monstera nechodomii	
		Monstera oblique	
		Monstera oblique 'Expilata' *(syn. Monstera epipremnoides)*	
		Monstera standleyana	
106.	Nandina	*Nandina domestica*	Nandinaceae
107.	-	*Neoregelia spectabilis*	Bromeliaceae
		*Neoregelia liberica (*syn. *Syngonium podophyllum)*	
108.	-	*Nephthytis liberica (*syn. *Syngonium podophyllum)*	
		Nephthytis liberica var. *variegata (*syn. *Syngonium angustatum* 'Albo lineatum'*)*	

Contd...

S. No.	Common Name	Botanical Name	Family
109.	-	*Muehlenbeckia platyclada (*syn. *Homalocladium platycladum)*	-
110.	-	*Musa wendlandiana (*syn. *Ensete ventricosum)*	-
111.	Nothopanax		
	-	*Nothopanax fruticosum (*syn. *Polyscias fruticosum)*	-
		*Nothopanax guifoylei (*syn. *Polyscias guilfoylei)*	
		*Nothopanax scutellarium (*syn. *Polyscias scutellaria)*	
112.	Lilyturf, Mondo grass	*Ophiopogon*	Liliaceae
	Jaburan lily turf, Snake beard, White lily turf	*Ophiopogon jaburan*	
	Variegated mondo grass	*Ophiopogon jaburan* 'Variegatus'	
	Dwarf lily turf, Mondo grass	*Ophiopogon japonicus*	
113.	Oplismenus		Poaceae
	Basket grass variegated	*Oplismenus burmanii (*syn. *Oplismenus hirtellus* 'Variegatus'*)*	
114.	Lady's sorrel, Wood-sorrel	*Oxalis*	Oxalidaceae
	Fire fern, Red flame	*Oxalis hedysaroides*	
115.	Cardinal's guard	*Pachystachys*	Acanthaceae
	Lollipop plant	*Pachystachys lutea*	
116.	Panax		
	-	*Panax balfourii (*syn. *Polyscias balfouriana)*	-
		*Panax excelsum (*syn. *Polyscias fruticosa)*	
		*Panax filicifolium (*syn. *Polyscias filicifolia)*	
		Panax fruticosa 'Elegans' *(*syn. *Polyscias fruticosa* 'Elegans'*)*	
		*Panax fruticosum (*syn. *Polyscias fruticosa)*	
117.	Screw pine	*Pandanus*	Pandanaceae
	Blue screw pine	*Pandanus baptistii*	
	-	*Pandanus baptistii* 'Aureus'	
		*Pandanus distichus (*syn. *Pandanus utilis)*	

Contd...

S. No.	Common Name	Botanical Name	Family
		Pandanus flabelliformis (syn. *Pandanus utilis*)	
		Pandanus heterocarpus	
		Pandanus mauritianus (syn. *Pandanus utilis*)	
	Bread fruit, Hala screw pine, Pandang, Pandanus palm, Thatch screw pine, Walking tree	*Pandanus odoratissimus*	
	-	*Pandanus odoratus* (syn. *Pandanus odaratissimus*)	
		Pandanus odorifer (syn. *Pandanus odaratissimus*)	
		Pandanus ornatus (syn. *Pandanus heterocarpus*)	
		Pandanus sanderi (*Pandanus tectorius* 'Roehsianus')	
		Pandanus tectorius (*Pandanus odoratissimus*)	
	Common screw pine	*Pandanus utilis*	
	Variegated screw pine, Veitch screw pine	*Pandanus veitchii*	
118.	-	*Panicum vaiegatum* (syn. *Oplismenus hirtellus* 'Variegatus')	Poaceae
119.	Pellionia		Urticaceae
	Trailing watermelon begonia	*Pellionia daveauana*	
	Rainbow vine, Satin Pellionia	*Pellionia pulchra*	
	-	*Pellionia repens*	
120.	Water melon begonia, Watermelon peperomia	*Peperomia arayreia*	Peperomiaceae
	-	*Peperomia arifolia* (syn. *P. argyreia*)	
		Peperomia clusaefolia (syn. *P. clusiifolia*)	

Contd...

S. No.	Common Name	Botanical Name	Family
	Red – edge peperomia	*Peperomia clusiifolia*	
	Ivy Peperomia, Ivy leaf peperomia, Platinum peperomia, Silver leaf peperomia	*Peperomia griseo-argentea*	
	-	*Peperomia hederafolia*	
		Peperomia hederifolia	
		Peperomia magnoliaefolia	
	Desert Privet	*Peperomia magnoliifolia*	
	-	*Peperomia metallica*	
	American Rubber plant, Baby Rubber plant, Pepper face	*Peperomia obtuifolia*	
	Gold tip	*Peperomia obtusifolia*	
		Peperomia pettifolia	
	Coin – leaf peperomia	*Peperomia polybotrya*	
	Philodendran peperomia	*Peperomia scandens*	
121.	-	*Perilepta dyerama* (syn. *Strobilanthes dyeranus*)	Acanthaceae
122.	Gardener's Garters, Ribbon grass	*Phalaris arundinaceae*	Poaceae
123.	Finger plant	*Philodendran bipinnatifidum*	Araceae
		Philodendron cordatum	
		Philodendran discolor	
		Philodendron dubia	
		Philodendron 'Emerald Queen'	
		Philodendron 'Florida Variegata'	
	Giant philodendron	*Philodendran giganteum*	
	-	*Philodendran guttiferum*	
		Philodendran lacerum	
		Philodendran laciniatum	
		Philodendran laciniosum	
		Philodendran microstictum	

Contd...

S. No.	Common Name	Botanical Name	Family
		Philodendran oxycardium	
		Philodendran pedatum	
		Philodendran pertusum	
		Philodendran pertusum 'Variegatum'	
		Philodendran pittieri	
		Philodendran scandens	
		Philodendran wendlandii	
124.	-	*Phrynium micholitzii*	Marantaceae
		Phrynium variegatum	
125.	-	*Phyllotaenium lindenii*	Araceae
126.	Aluminium plant, Friendship plant,	*Pilea cadierei*	Urticaceae
	Artillery plant	*Pilea microphylla*	
		Pilea mucosa	
127.	-	*Pitcairnia heterophylla*	Bromeliaceae
		Pitcairnia morrenii	
128.	-	*Pleomele reflexa*	Asparagaceae
		Pleomele reflexa 'Angustifolia'	
		Pleomele reflexa 'Variegata'	
		Pleomele thalioides	
129.	Balfour Aralia, Dinner Plate Aralia	*Polyscias balfouriana*	Araliaceae
	-	*Polyscias balfouriana* 'Marginata'	
	White aralia	*Polyscias balfouriana* 'Pennockii'	
130.	Fern – leaf Aralia	*Polyscias filicifolia*	
	Ming Aralia, Parsley Panax	*Polyscias fruticosa*	
		Polyscias fruticosa 'Elegans'	
	Coffee tree. Gerium-leaf Aralia	*Polyscias guilfoylei*	
	-	*Polyscias guilfoylei* 'Crispa'	
	-	*Polyscias paniculata* 'Variegata'	
	Saucer Panax	*Polyscias scutellaria*	
131.	-	*Pothos aureus*	Araceace
		Pothos cannaefolia	
		Pothos jambea	
		Pothos wilcoxii	

Contd...

S. No.	Common Name	Botanical Name	Family
132.	-	*Pseuderanthemum atropupureum*	Acanthaceae
		Pseuderanthemum atropupureum 'Tonga'	
	Purple False Eranthemum	*Pseuderanthemum atropupureum* 'Tricolor'	
	-	*Pseuderanthemum bicolor*	
		Pseuderanthemum laxiflorum	
		Pseuderanthemum reticulatum	
		Pseuderanthemum sinuatum	
133.	-	*Puya heterophylla*	Bromeliaceae
134.	-	*Pyrrheima fuscata*	Commelinaceae
		Pyrrheima loddigesii	
135.	-	*Raphidophora aurea*	Araceae
		Raphidophora pinnata	
136.	Traveler's Palm	*Ravenala madagascariensis*	Strelitziaceae
137.	Lady Palm	*Raphis excelsa*	Araceae
	-	*Raphis flabelliformis*	
		Raphis humile	
	Reed Rhapis	*Raphis humilis*	
	-	*Raphis kwanwortsick*	
138.	-	*Rhektophyllum mirabile*	Araceae
139.	-	*Rhoeo discolor*	Commelinaceae
	Boat lily, Man-in-a-boat, Moses-in-a-boat, purple leaved spider wort, Two-men-in-a-boat	*Rhoeo spathacea*	
	Variegated Boat lily	*Rhoeo spathacea* 'Variegata'	
	-	*Rhoeo spathacea* 'Vittata'	
140.	-	*Rhynchophorum obtusifolium*	Curculionidae
141.	Mouse Thorn	*Ruscus hypoglossum*	Ruscaceae
142.	-	*Sanchezia glaucophylla*	Acanthaceae
		Sanchezia nobilis	
		Sanchezia speciosa	
		Sanchezia spectabilis	
	Snake plant, Devil's Tongue, Lucky plant, Hemp plant, Bowstring - Hemp	*Sansevieria craigii*	Agavaceae

Contd...

S. No.	Common Name	Botanical Name	Family
	-	*Sansevieria cylindrical*	
		Sansevieria fragrans	
		Sansevieria laurentii	
		Sansevieria metallica	
		Sansevieria sulcata	
	Mother – law-tongue, Snake plant	*Sansevieria trifasciata*	
	-	*Sansevieria trifasciata* 'Craigii'	
		Sansevieria trifasciata 'Golden Hahnii'	
	Bird's – nest sansevieria	*Sansevieria trifasciata* 'Hahnii'	
	-	*Sansevieria trifasciata* 'Laurentii'	
		Sansevieria trifasciata 'Silver Hahnii'	
	Ceylon Bow-string Hemp	*Sansevieria zeylanica*	
143.	Prince's Feather	*Saxifraga chinensis*	Saxifragaceae
	-	*Saxifraga japonica*	
		Saxifraga sarmentosa	
	Beefsteak Geranium, Creeping Sailor, Loving Sailor, Mother – of – thousands, Strawberry begonia	*Saxifraga stolonifera*	
144.	Umbrella Tree, Rubber tree, Star-leaf Patete	*Schefflera actinophylla*	Araliaceae
	-	*Schefflera macrostachya*	
		Schefflera venulosa	
145.	Drop-tongue	*Schismatoglottis neoguineensis*	Araceae
		Schismatoglottis novoguinensis	
		Schismatoglottis roebelinii	
		Schismatoglottis variegata	
		Schizocasia portei	
		Schizocasia regnieri	
146.	-	*Scindapsus aureus*	Araceae
		Scindapsus aureus 'Marble Queen'	
		Scindapsus cuscuaria	
	Silver vine	*Scindapsus pictus*	

Contd...

S. No.	Common Name	Botanical Name	Family
147.	Purple Heart	*Setcreasea pallida*	Commelinaceae
		Setcreasea purpurea	
		Setcreasea striata	
		Setcreasea tampicana	
148.	Brown spider wort	*Siderasis fuscata*	Commelinaceae
149.	-	*Spironema dracaenoides*	
	-	*Spironema fragrans*	
150.	-	*Stenotaphrum americanum* 'Variegatum'	Poaceae
	Buffalo Grass, St. Angustine Grass	*Stenotaphrum secundatum* 'Variegatum'	
151.	-	*Strelitzia parvifolia*	Strelitziaceae
	Bird –of - paradise	*Strelitzia reginae*	
	Persian shield	*Strobilanthes dyerianus*	Acanthaceae
152.	-	*Stromanthe humilis*	Marantaceae
		Stromanthe sanginea	
153.	-	*Syngonium albo-lineatum*	Araceae
		Syngonium angustatum	
	African Evergreen, Arrow-head vine	*Syngonium podophyllum*	
	-	*Syngonium podophyllum* var. *albo-lineatum*	
		Syngonium podophyllum 'Ruth Fraser'	
		Syngonium podophyllum 'Tricolor'	
154.	-	*Taetsia fruticosa*	Asparagaceae
155.	-	*Talinum paniculatum*	Portulacaceae
		Talinum patens	
156.	-	*Terminalia elegans* 'Variegata'	Combretaceae
157.	-	*Thalia sanguine*	Marantaceae
		(syn. *Stromanthe sanguinea*)	
158.	-	*Toenelia fragrans*	Araceae
		(syn. *Monstera deliciosa*)	
159.	-	*Trachycarpus excelsa*	Arecaceae
160.	Spiderwort	*Tradescantia albiflora*	Commelinaceae
	Giant white Inch plant	*Tradescantia albiflora* 'Albo-vittata'	
	Wandering Jew	*Tradescantia albiflora* 'Variegata'	
	-	*Tradescantia bicolor*	
		Tradescantia discolor	

Contd...

S. No.	Common Name	Botanical Name	Family
		Tradescantia discolor 'Variegata'	
		Tradescantia dracaenoides	
		Tradescantia fluminensis	
		Tradescantia fuscata	
		Tradescantia multicolor	
	Chain plant	*Tradescantia navicularis*	
	-	*Tradescantia pexata*	
		Tradescantia quadricolor	
		Tradescantia reginae	
	White Gossamer, White velvet, White velvet creeper	*Tradescantia sillamontana*	
	-	*Tradescantia spathacea*	
		Tradescantia striata	
		Tradescantia varigata	
		Tradescantia tricolor	
		Tradescantia viridis	
		Tradescantia zebrina	
161.	Vanilla	*Vanilla aromatica*	Orchidaceae
		Vanilla fragrans	
		Vanilla planifolia	
162.	Spoon flower	*Xanthosoma lindenii*	Araceae
	-	*Xanthosoma mafaffa*	
	Blue Ape, Blue Taro, Violet-Stemmed Taro	*Xanthosoma violaceum*	
163.	-	*Zamia villosa*	Zamiaceae
164.	Inch Plant, Silvery Wondering Jew	*Zebrina pendulla*	Commelinaceae
	Happy wandering Jew	*Zebrina pendula* 'Quadricolor'	

Indoor Plants

Acalypha hispida

Acalypha wilkesiana 'Forma Circinata'

Acalypha wilkesiana

Acalypha wilkesiana 'Macrophylla'

Aglaonema brevispathum f. hospitum

Aglaonema commutatum

Aglaonema costatum

Aglaonema crispum

Aglaonema hookerianum

Aglaonema modestum Variegatum

Aglaonema nitidum

Alocasia amazonica

Alocasia brancifolia

Alocasia 'Calidora'

Alocasia cucullata

Alocasia 'Frydek'

Alocasia 'Hilo Beauty'

Alocasia infernalis 'Kapit'

Alocasia lauterbachiana

Alocasia macrorrhiza

Alocasia macrorrhiza 'Black Stem'

Alocasia macrorrhiza 'Borneo Giant'

Alocasia macrorrhiza 'Lutea'

Alocasia 'Mandalay'

Alocasia odora

Alocasia plumbea nigra

Alocasia 'Polly'

Alocasia portei

Alocasia princeps 'Purple Cloak'

Alocasia reginula 'Black Velvet'

Alocasia sanderiana 'Nobilis'

Alocasia wentii 'Variegata'

Alpinia 'Sanderae'

Alpinia zerumbet 'Variegata'

Ananas comosus 'Variegatus'

Anthurium bakeri

Anthurium brownii

Anthurium clarinervium

Anthurium cordatum

Anthurium faustmirandae

Anthurium podophyllum

Anthurium pseudospectabile

Anthurium veitchii

Anthurium watermaliense

Asparagus densiflorus 'Myers'

Asparagus densiflorus 'Sprengeri'

Asparagus setaceus 'Pyramidalis'

Aspidistra elatior 'Asahi'

Aspidistra elatior

Aspidistra elatior 'Variegata'

Aspidistra 'Mary Sizemore'

Aspidistra 'Milky Way'

Begonia heracleifolia

Begonia ramentacea

Begonia rex 'Silver Queen'

Begonia speculata

Beloperone guttata

Beloperone guttata 'Yellow Queen'

Billbergia pyramidalis var. 'Concolor'

Brassaia actinophylla

Caladium 'Blaze'

Caladium 'Carolyn Wharton'

Caladium 'Fire Chief'

Caladium 'Florida Elise'

Caladium 'Florida Sweetheart'

Caladium 'Harlequin'

Caladium humboldtii 'Mini White'

Caladium humboldtii 'Myriostigma'

Caladium 'Kathleen'

Caladium 'Lance Wharton'

Caladium lindenii 'Magnificum'

Caladium 'Pink Beauty'

Caladium 'Red Flash'

Caladium 'Red Frill'

Caladium 'Hue of Pink'

Calathea bachemiana

Calathea bella

Calathea exotica

Calathea fasciata

Calathea 'Helen Kennedy'

Calathea lietzei

Calathea mediopicta

Calathea ornata

Calathea princeps

Calathea undulata

Calathea zebrina

Callisia elegans

Callisia fragrans

Calathea rufibarba

Chlorophytum comosum 'Mandaianum'

Chlorophytum comosum 'Vittatum'

Chrysalidocarpus lutescens

Codiaeum variegatum

Cordyline terminalis

Cordyline terminalis 'Madame Eugene Andre'

Costus speciosus

Ctenanthe oppenheimiana

Curculigo capitulata

Cycas circinalis

Cycas revoluta

Cyperus albostriatus

Cyperus alternifolius 'Variegatus'

Cyrtosperma johnstonii

Dichorisandra reginae

Diefenbachia amoena

Diefenbachia amoena 'Tropic Snow'

Dieffenbachia maculata

Dieffenbachia oerstedii

Dieffenbachia picta

Dieffenbachia 'Exotica Alba'

Dizygotheca elegantissima

Dorstenia turneraefolia

Dracaena deremensis 'Warneckii'

Dracaena fragrans

Dracaena fragrans 'Massangeana'

Dracaena fragrans 'Victoria'

Dracaena goldieana

Dracaena reflexa 'Variegata'

Dracaena surculosa

Dracaena surculosa 'Florida Beauty'

Dracaena thalioides

Dracaena umbraculifera

Dracaena surculosa 'Milky Way'

Drimiopsis kirkii

Ensete ventricosum

Epipremnum aureum

Epipremnum aureum 'Marble queen'

Epipremnum pinnatum

Eranthemum nigrum

Euonymus japonicus

Excoecaria bicolor

Fatsia japonica

Fatsia japonica 'Spiders Web'

Ficus elastica 'Decora'

Ficus elastica 'Variegata'

Ficus lyrata

Ficus pumila

Fittonia verschaffeltii var. *Argyroneura*

Geogenanthus undatus

Gloxinia perennis

Graptophyllum pictum

Graptophyllum pictum 'Alba Variegata'

Gymnostachyum ceylanicum

Gynura aurantiaca

Gynura procumbens

Hedychium coronarium

Hemigraphis alternata

Hemigraphis 'Exotica'

Hippobroma longiflora

Hoffmannia refulgens

Homacladium platycladum

Homalomena rubescens

Hypoestes phyllostachya

Hoya carnosa

Hoya carnosa 'Variegata'

Iresine herbstii

Kaempferia gilbertii

Kaempferia roscoeana

Kaempferia rotunda

Liriope muscari

Livistona chinensis

Manihot esculenta 'Variegata'

Maranta arundinacea

Maranta arundinacea 'Variegata'

Maranta leuconeura

Maranta 'Repens'

Monstera deliciosa

Monstera deliciosa 'Marmorata'

Monstera epipremnoides

Monstera obliqua

Nandina domestica

Neoregelia spectabilis

Ophiopogon jaburan

Ophiopogon japonicus

Ophiopogon japonicus 'Variegatus'

Oplismenus hirtellus 'Variegatus'

Oxalis hedysaroides 'Rubra'

Pachystachys lutea

Pandanus tectorius

Pandanus tectorius 'Baptistii'

Pandanus veitchii

Pellionia daveauana

Pellionia pulchra

Peperomia argyreia

Peperomia caperata

Peperomia clusiifolia

Peperomia griseo-argentea

Peperomia obtusifolia 'Golden Genua'

Peperomia obtusifolia

Peperomia obtusifolia 'Variegata'

Peperomia polybotrya 'Jayde'

Phalaris arundinacea var. 'Picta'

Philodendron bipinnatifidum

Philodendron discolor

Philodendron Florida Beauty Alba

Philodendron lacerum

Philodendron pedatum

Philodendron scandens oxycardium

Philodendron xanadu

Pilea cadierei

Pilea microphylla

Polyscias balfouriana 'Marginata'

Polyscias balfouriana 'Pennockii'

Polyscias filicifolia

Polyscias paniculata variegata

Polyscias balfouriana

Pseuderanthemum atropurpureum

Pseuderanthemum atropurpureum rubrum

Pseuderanthemum atropurpureum tricolor

Pseuderanthemum atropurpureum 'Variegated'

Pseuderanthemum atropurpureum
'Albomarginatum'

Pseuderanthemum carruthersii var.
atropurpureum 'Variegatum'

Pseuderanthemum laxiflorum

Pseuderanthemum nigrum

Pseuderanthemum reticulatum

Pseuderanthemum sinuatum

Ravenala madagascariensis

Rhapis excelsa

Rhapis humilis

Rhoeo discolor 'Variegata'

Rhoeo spathacea 'Vittata'

Ruscus hypoglossum

Sanchezia speciosa

Sanchezia trifasciata 'Golden Hahnii'

Sanchezia trifasciata 'Hahnii'

Sanseveria trifasciata 'Silver Hahnii'

Sansevieria trifasciata Laurenti

Sansevieria cylindrica

Saxifraga stolonifera

Schismatoglottis neoguineensis

Scindapsus pictus

Senecio herreanus

Setcreasea pallida 'purple heart'

Siderasis fuscata

Strobilanthes dyerianus 'Persian Shield'

Stromanthe sanguinea

Syngonium angustatum 'Albolineatum'

Talinum paniculatum

Tradescantia albiflora 'Albo-vittata'

Tradescantia navicularis

Tradescantia sillamontana

Tradescantia spathacea 'Variegata'

Tradescantia albiflora

Vanilla planifolia

Vanilla planifolia 'Variegata'

Xanthosoma lindenii

Zebrina pendula

Zebrina pendula 'Quadricolor'

Annuals

Cosmos bipinnatus

Gaillardia pulchella

Pelargonium graveolens

Petunia hybrida

Portulaca grandiflora

Tagetes erecta

Verbena hybrida

Zinnia elegans

Bulbous Plants

Canna indica

Gerbera jamesonii

Gloriosa superba

Hedychium coronarium

Nymphaea spp.

Polianthes tuberosa

Cacti

Astrophytum myriostigma

Cephalocereus senilis

Chamaecereus silvestrii

Echinocactus grusonii

Ferocactus latispinus

Gymnocalycium mihanovichii

Lophophora williamsii

Mammillaria bocasana

Melocactus concinnus

Notocactus graessneri

Notocactus leninghausii

Opuntia microdasys

Climbers

Antigonon leptopus

Aristolochia grandiflora

Artabotrys odoratissimus

Campsis grandiflora

Clematis gouriana

Clerodendrum splendens

Quisqualis indica

Thunbergia laurifolia

Ferns

Adiantum venustum

Dicksonia antarctica

Dryopteris-filix-mas

Matteuccia struthiopteris

Microlepia strigosa

Polypodium vulgare

Polystichum aculeatum

Polystichum setiferum

Selaginella strigosa

Shrubs

Allamanda cathartica

Allamanda purpurea

Bougainvillea

Calliandra brevipes

Cestrum nocturnum

Gardenia jasminoides

Hibiscus rosa sinensis

Ixora chinensis

Ixora lutea

Lagerstroemia indica

Murraya paniculata

Mussaenda erythrophylla

Nerium indicum

Pentas

Russelia juncea

Tabernaemontana coronaria

Trees

Amherstia nobilis

Anthrocephalus cadamba

Bauhinia variegata

Couroupita guianensis

Jacaranda mimosaefolia

Mimusops elengi

Magnolia grandiflora

Mesua ferrea

Michelia champaca

Millingtonia hortensis

Nyctanthes arbor tristis

Plumeria acuminata

Plumeria acutifolia

Plumeria rubra

Pterospermum acerifolium

Saraca asoca

Tabebuia pallida

Tabebuia rosea

Succulents

Agave americana

Agave filifera

Agave stricta

Agave Victoria – reginae

Aloe ferox

Aloe humilis

Anacampseros alstonii

Anacampseros papyracea

Anacampseros rufescens

Anacampseros ustulata

Aptenia cordifolia

Bergeranthus scapigera

Caralluma burchardii

Cotyledon undulata

Crassula arborescens

Crassula falcata

Crassula perfoliata

Duvalia elegans

Dyckia sulphurea

Echinopsis dammaniana

Euphorbia bupleurifolia

Euphorbia grandidens

Euphorbia horrida

Euphorbia ingens

Euphorbia obesa

Euphorbia polygona

Euphorbia valida

Greenovia aurea

Haworthia coarctata

*Haworthia reinwardhii
chalwinii*

Haworthia reinwardtii

Kalanchoe beharensis

Kalanchoe marmorata

Kalanchoe tomentosa

Portulacaria afra

www.ingramcontent.com/pod-product-compliance
Lightning Source LLC
Chambersburg PA
CBHW050518190326
41458CB00005B/1577